T0303894

Mineralogy for Petrologists

Mineralogy for Petrologists

Optics, chemistry and occurrence of rock-forming minerals

Michel Demange
Centre de Geosciences, École des Mines
Paris, France

Fluid and melt inclusions in rock-forming minerals (pp. 44–49) by Jacques Touret, Ecole de Mines Paris, France

CRC Press
Taylor & Francis Group
Boca Raton London New York Leiden

CRC Press is an imprint of the
Taylor & Francis Group, an **informa** business

A BALKEMA BOOK

CRC Press/Balkema is an imprint of the Taylor & Francis Group, an informa business

© 2012 Taylor & Francis Group, London, UK

Typeset by V Publishing Solutions Pvt Ltd., Chennai, India
Printed and bound by CPI Group (UK) Ltd, Croydon, CR0 4YY

Published by: CRC Press/Balkema
          P.O. Box 447, 2300 AK Leiden, The Netherlands
          e-mail: Pub.NL@taylorandfrancis.com
          www.crcpress.com – www.taylorandfrancis.com

Library of Congress Cataloging-in-Publication Data

Demange, Michel (Michel Andre)
    Mineralogy for petrologists : optics, chemistry, and occurrences
    of rock-forming minerals / Michel Demange.
       p. cm.
    Summary:"This book provides a categorized and visualized
    overview and presents microscopic observations, systematic
    mineralogy, chemistry, geology, stability, paragenesis, occurrence
    and use in petrology of 137 minerals. Structural formula
    calculations are included in the appendix. Consists of a set
    of book and CD-ROM for students and practically-oriented
    researchers and professionals in geology, geological, mining,
    and mineral resources engineering who need a reference
    of mineralogy, applied to petrology. The CD-ROM contains
    384 color plates with mineral microscopic visuals under
    various circumstances"-- Provided by publisher.
    Includes bibliographical references and index.
    ISBN 978-0-415-68421-7
    1. Petrology. 2. Minerals. I. Title.

    QE431.2.D46 2012
    549--dc23
                                         2012007518

ISBN: 978-0-415-68421-7 (Hbk)

Contents

Author biography

Dr. Michel Demange has devoted his career to regional geology and tectonics of metamorphic and magmatic terranes and to ore deposits. Graduated from the École Nationale Supérieure des Mines de Paris and holding a Docteur-es-Sciences of the University Pierre et Marie Curie, Paris VI, he has been active in a rich variety of geological projects and investigations around the world. In combination with his teaching and research activities at the École des Mines in Paris, France, he headed various research studies. This book benefits from the great experience in field studies, research and teaching and the wealth of data and images accumulated during his career. It will be followed by the companion volume "Rock Textures: Igneous, Metamorphic rocks and Deformation Textures (CRC Press/Balkema). Both volumes are available in French from the Presses de l'Ecole des Mines, Paris.

Introduction

WHY THE MICROSCOPE? PURPOSE OF THE BOOK

The purpose of **petrology** is to understand the conditions of the formation of rocks. The first stage of this approach is to describe and classify the rocks; that is the subject of **petrography**. Many characters can be used: density, hardness, colour, structure (on the scale of the outcrop), texture (on the scale of the sample or of the microscope), mineralogical composition, chemical composition, speed of propagation of the waves, etc. Among all the characters, the mineralogical composition is certainly the most important: it allows us to, more or less, accurately foresee the other characters and it can be used to determine the very conditions of the formation of rocks. Indeed, the nature and chemical composition of the minerals in a rock obey stricter laws, since the formation temperature was higher.

Two methods of determining the minerals are relatively simple, fast and inexpensive: the X-ray diffraction and petrographic microscope.

The X-ray diffraction allows precise identification of minerals, but generally requires a separation. This method remains absolutely necessary in the case of very fine grained rocks and minerals. However, its use can be difficult when the rock contains numerous minerals. Minerals in a small proportion may remain unnoticed. And above all, this method does not allow us to observe and discuss the textures of the rock and the relations between its various minerals.

The use of the petrographic microscope overcomes these difficulties – at least in the case of rocks whose grain is large enough (typically greater than the thickness of the thin section, that is to say 30 microns).

The study of a more or less important number of thin sections remains a prerequisite for more complex and more expensive methods, such as chemical analysis of individual minerals, either by wet chemistry or by microprobe.

However, the petrographic microscope requires some learning, and some experience. There are books and tables, but it is rare that one comes to determine a completely unknown mineral using only tables.

The purpose of this work is to illustrate the most common rock forming minerals as they appear under the petrographic microscope and facilitate their learning. The selected mineral species have been chosen in reference to the classical book by Deer, Howie and Zussman, *An Introduction to the Rock Forming Minerals* (first edition 1966). This guide includes two parts, a book and a CD.

The CD illustrates the microscopic appearance of the rock forming minerals and shows numerous thin sections in plane and cross polarized light. Each mineral is shown by cards containing its name, its chemical composition, some sketches showing its forms, its various optical characters and several photomicrographs illustrating these characters. There are generally two photomicrographs taken in the same position, one in plane polarized light, the other in cross polarized light. The colors on a computer screen, without reaching the luminosity of the microscope, are certainly superior to printed photographs. A CD has the advantage of showing in a reduced format numerous illustrations so that a same mineral can be presented in several cards showing its various habits and various occurrences.

The book is deliberately brief, and develops the issues for which the illustrations on the CD are of less interest. The first part deals with the definition of a mineral species and the factors of its occurrence. Then the methods of microscopic observation are quickly described. The bulk of the book is a series of monographs on different minerals or mineral groups. This book does not pretend to replace extensive treaties of mineralogy but insists on the more important points for the characterization of a mineral:

- some data on the **structure** of the mineral, since it explains its chemical formula.
- the **chemical composition** and the variations of composition that can be expected in the same mineral or group of minerals as well in major elements as in some minor elements.
- **conditions of stability**, chemical composition and stability conditions lead to discussions of the **various occurrences** of the mineral: given that mineral is not found in just any rock; the various paragenesis (mineral association at equilibrium) do not occur arbitrarily and changes in chemical composition of a mineral report on various geological evolutions. This concern for occurrences and geological evolution determines the plan adopted in this guide, that is somewhat different from the classical mineralogical classification. The principles of the calculation of structural formulas are given in Annex.
- the optical characters of the most important minerals are shown in the book by deliberately concise summary tables.

Use of the CD

A mineral may be more or less rare or common, as it is indicated on the cards: VC = very common, C = common, RC = rather common, QR = quite rare, R = rare.

It depends, of course, on the rocks or problems studied: wollastonite is fairly common in metamorphic marbles, very rare in granites. Similarly, some habits may be much rarer than the common habits of the same mineral: euhedral quartz is fairly infrequent, whereas quartz is a very common mineral.

Browsing the CD

One can navigate backwards and forwards from one card to another by using the mouse to click on the ◀ ▶ symbols or by using the left/right arrow keys on the keyboard.

- Sign 🏠 takes you back to the general index
- Sign 🏠 takes you back to the index of each part
- Sign ↻ takes you back to the last card shown. Please double click on it.
- Click on the images to see the larger image in a pop-up window
- All dark blue text is clickable and will direct you to a particular card
- The ● sign adds additional linking to related texts

Chapter 1

Rocks and minerals

1.1 WHAT IS A MINERAL?

A mineral is a **naturally occurring** homogeneous solid characterized by a highly **ordered atomic structure** and characteristic **chemical composition**.

The International Mineralogical Association gives the following definition: "a mineral is an element or chemical compound that is normally crystalline and that has been formed as a result of geological processes" (Nickel, 1995). This definition excludes synthetic "minerals".

1.1.1 An ordered atomic structure

Minerals are solids made of atoms arranged in a periodic and symmetric lattice. Frankenheimer (1842) then Bravais (1848) have shown that there are 14 (and only 14) basic crystal lattice arrangements of atoms in three dimension types, referred to as the "Bravais lattice". These lattices derive from seven basic reticular systems: **triclinic, monoclinic, orthorhombic, tetragonal, trigonal, hexagonal, cubic.** These systems are characterized by elements of symmetry: centers, plans, reverse plans (symmetry with respect to a plane and rotation), axes (order 2, 3, 4 or 6), reverse axes (rotation and symmetry about center) (Figure 1.1).

The unit cell is the smallest crystal volume which has all the geometric properties (symmetry, size), physical and chemical properties of crystal. It is defined by the lengths of three vectors a, b, c and three angles α, β, γ.

These three vectors form a basis, in which are identified in any plane or any vector. In this basis, the equation of a plane is:

u x/y + v y/b + w z/c = 1

u, v and w are integers and are called **Miller indices**. The notation of a crystal face or a plane is (h k l) (round brackets). By convention the notation of a negative index (−u) is (ū). The vector normal to this plane has u, v, w coordinates; so the notation of such direction is [uvw] (square brackets).

Hexagonal and trigonal systems use a set of coordinates and notations slightly different: 3 axis at 120° (x y t) are used in the (x y) plane; the z axis is

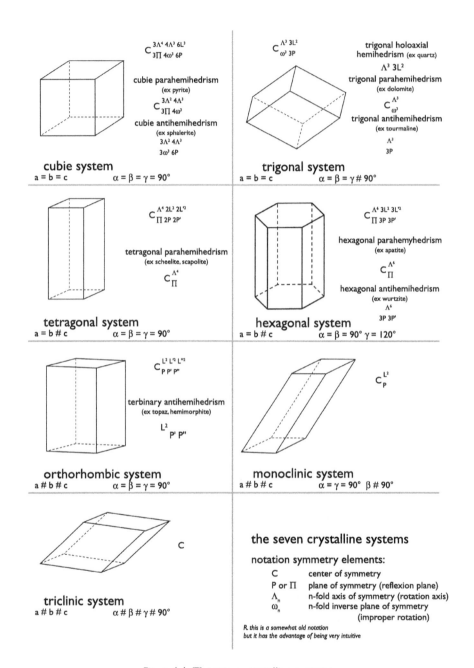

$C\,^{3\Lambda^4\,4\Lambda^3\,6L^3}_{3\Pi\,4\omega^3\,6P}$

cubie parahemihedrism
(ex pyrite)

$C\,^{3\Lambda^2\,4\Lambda^3}_{3\Pi\,4\omega^3}$

cubie antihemihedrism
(ex sphalerite)

3Λ² 4Λ³
3ω³ 6P

cubie system
a = b = c   α = β = γ = 90°

$C\,^{\Lambda^3\,3L^2}_{\omega^3\,3P}$

trigonal holoaxial
hemihedrism (ex quartz)

Λ³ 3L²

trigonal parahemihedrism
(ex dolomite)

$C\,^{\Lambda^3}_{\omega^3}$

trigonal antihemihedrism
(ex tourmaline)

Λ³
3P

trigonal system
a = b = c   α = β = γ ≠ 90°

$C\,^{\Lambda^4\,2L^2\,2L'^2}_{\Pi\,2P\,2P'}$

tetragonal parahemihedrism
(ex scheelite, scapolite)

$C\,^{\Lambda^4}_{\Pi}$

tetragonal system
a = b ≠ c   α = β = γ = 90°

$C\,^{\Lambda^6\,3L^2\,3L'^2}_{\Pi\,3P\,3P'}$

hexagonal parahemyhedrism
(ex apatite)

$C\,^{\Lambda^6}_{\Pi}$

hexagonal antihemihedrism
(ex wurtzite)

Λ⁶
3P 3P'

hexagonal system
a = b ≠ c   α = β = 90° γ = 120°

$C\,^{L^2\,L'^2\,L''^2}_{P\,P'\,P''}$

terbinary antihemihedrism
(ex topaz, hemimorphite)

L²
P' P''

orthorhombic system
a ≠ b ≠ c   α = β = γ = 90°

$C\,^{L^2}_{P}$

monoclinic system
a ≠ b ≠ c   α = γ = 90° β ≠ 90°

C

triclinic system
a ≠ b ≠ c   α ≠ β ≠ γ ≠ 90°

the seven crystalline systems

notation symmetry elements:

C	center of symmetry
P or Π	plane of symmetry (reflexion plane)
Λₙ	n-fold axis of symmetry (rotation axis)
ωₙ	n-fold inverse plane of symmetry (improper rotation)

R. this is a somewhat old notation
but it has the advantage of being very intuitive

Figure 1.1 The seven crystalline systems.

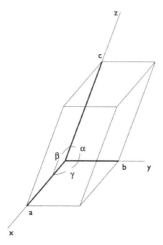

Figure 1.2 The unit cell.

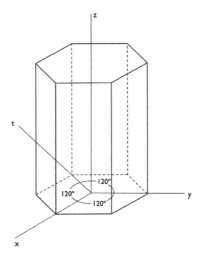

Figure 1.3 Set of coordinates in the hexagonal
and trigonal systems.

perpendicular to the plane (x y). The notation of a plane in these coordinates
will be (j h k l), j, h and k along the axes x y t and l along the z axis, with:

$$j + h + k = 0$$

When the lattice and the crystal have the same symmetry, which is then
at maximum, it is called holohedry.

When the symmetry of the crystal is lower than the one of the lattice, one talks of merohedrism (hemihedrism if it contains only half of the faces of the corresponding, tetartohedrism with a quarter of the faces, etc.). The most common merohedrism are:

- *antihemihedrism:* absence of center (and of the associated planes); an example of antihémihedrism is the (cubic or tetragonal) tetrahedron that derives from the octahedron; another example is the tourmaline prisms which have two extremities that are not symmetrical; such minerals have piezoelectric properties;
- *parahemihedrism:* the center of symmetry is present, but the binary axes are absent; for instance, the pentagonal dodecahedron (a common form of pyrite, or pyritohedron) derives from an hexatetrahedron;
- *holoaxial hemihedrism:* no center, no binary axes; trigonal quartz and scheelite belong to this class.

Any lattice plane can form a face of a crystal. But growth is more or less rapid according to the directions. The fast growing faces are rapidly eliminated and slow growing faces become dominant. These are the faces that determine the **crystalline forms** of the mineral. The crystalline forms reflect the structure of the lattice and the symmetry system to which it belongs. When a mineral has its own crystal forms it is said to be euhedral, otherwise we talk of an anhedral mineral. In rocks, minerals such as tourmaline, kyanite and garnet are frequently euhedral, others, such as quartz, more rarely.

A **cleavage** is a perfect and repetitive separation plane. The cleavage planes reflect, at the macroscopic scale, one or more zones of weakness: cleavage typically occurs preferentially parallel to higher density planes. The cleavages show the same symmetry as the crystal. Some minerals have no cleavage (quartz, for example), others have one or more.

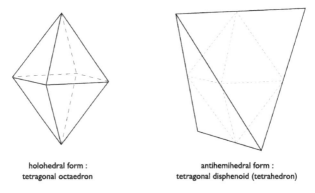

holohedral form :
tetragonal octaedron

antihemihedral form :
tetragonal disphenoid (tetrahedron)

Figure 1.4 An example of antihemihedral form.

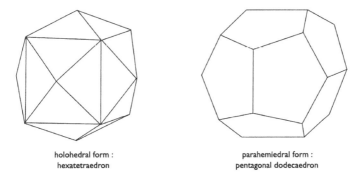

holohedral form :   parahemiedral form :
hexatetraedron   pentagonal dodecaedron

Figure 1.5 An example of parahemihedral form.

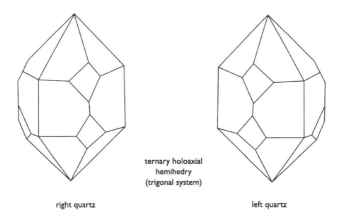

ternary holoaxial
hemihedry
(trigonal system)

right quartz   left quartz

Figure 1.6 An example of holoaxial hemihedry.

A **fracture** is a more or less irregular surface, not repeated or repeated a few times. The appearance of some fractures may be characteristic of a given mineral. An example is the tourmaline with fractures transversal on the prism.

A **twin** is an intergrowth of two or several crystals of the same mineral species, joined by a determined symmetry law. There are two types of twins:

1   **Contact twins:** the two individuals are symmetric to a plane, the twinning plane; the surface of accolement is then a plane (example: twinning of albite in plagioclases);
2   **Penetration twins:** the two individuals are symmetric to an axis, the twinning axis, the accolement surface is then *any* (example: Carlsbad twinning in feldspars).

Twins may by simple or repeated. Among the repeated twins there are:

- *Polysynthetic twins*: the joined individual crystals form parallel lamellae (examples: twins in cordiérite; albite and pericline twins in the plagioclases, grunerite).
- *Cyclic twins*: individual crystals form a more or less circular association (*example*: twins of leucite, cordierite).

The unit cell of the twinned system has a higher symmetry than the unit cell of the individual crystals.

1.1.2 A given chemical composition

A given chemical composition is not sufficient to define a mineral. Indeed **polymorphs** are minerals with the same chemical composition but a different structure: diamond (cubic) and graphite (hexagonal), calcite (trigonal) and aragonite (orthorhombic) are well known examples. Other examples are given by the various polymorphs of silica or feldspar.

The chemical composition varies within certain limits depending on:

- the presence of trace elements included in the lattice;
- substitutions between elements:
  - such substitutions can be made of atom to atom, for example, the substitution $Mg \Leftrightarrow Fe^{2+}$;
  - or by laws of substitution involving several elements. An example is the widespread substitution in silicates

$$Si^{IV} Mg^{VI} \Leftrightarrow Al^{IV} Al^{VI}$$

(where Al^{IV} is aluminum in 4-folds coordination, in tetrahedral site, surrounded by 4 atoms of oxygen and Al^{VI} aluminum in 6-fold coordination, and octahedral site, surrounded by 6 atoms of oxygen).

Micas and amphibole groups are minerals where the laws of substitution are particularly diverse.

The substitutions in minerals are governed by strict laws (rules of Goldschmidt):

- electrical equilibrium must be respected: for instance in the above example the valency of silicon is 4, the one of the magnesium is 2: $4 + 2 = 6$, the valency of aluminum is 3: $3 + 3 = 6$;
- only elements that have ionic radii difference of less than 15% can replace each other; the difference may be bigger but the substitution is then only partial. Iron replaces magnesium in common silicates and carbonates, but the replacement of magnesium by calcium is impossible (or at least very limited) in carbonates due to the difference in ionic

radii of these elements. When two elements have the same charge, the one that has the smallest ionic radius is preferentially incorporated: so, in an isomorphous series, the magnesium-bearing members are stable at higher temperatures than the iron-bearing members;

• in the lattice, a site may remain all or partially empty; for instance, the substitution 2 Al^{VI} □ ⇔ 3 (Fe, Mg)^{VI}, (where the sign □ represents a vacant site), is (limited) substitution occurring between dioctahedral micas and trioctalhedral micas.

The composition of minerals is even more constrained as the temperature is higher. Knowledge of the precise compositions of minerals and their variations is very important to reconstruct the conditions of the formation of rocks.

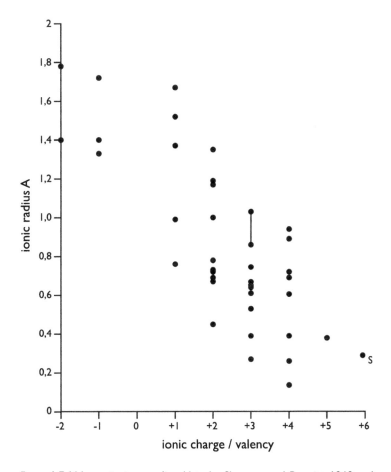

Figure 1.7 Valency, ionique radius (data by Shannon and Prewitt, 1969 and Shannon, 1976).

Optical methods allow certain precise determinations. The founder of these methods was undoubtedly A. Michel-Levy with his many charts for feldspars (1894–1904). The books by Roubault et al. *Determination of minerals of the rocks with the petrograhic microscope* (1963, Fabriès et al. 1984) and Tröger *Optische Bestimmung des gesteinsbildenden Minerale* (1971), provides numerous charts to determine with the microscope, or at least to approach the composition of many minerals. These methods are simple and cheap, but they provide only approximations. They are still widely used for rapid determination of the plagioclases. The invention of the electron microprobe analysis in 1951 allows almost punctual measures. The microprobe is now a common tool in all laboratories.

When studying a given geological phenomenon (a magmatic series, a prograde metamorphism or a metasomatic zoning, for instance), it commonly appears that the chemical composition of a given mineral varies more or less regularly through the various stages of this phenomenon. There are various laws of substitution playing simultaneously: for instance, a biotite is simultaneously enriched iron (by the substitution $Mg \geq Fe$) and aluminum (according the substitution $Si\ Mg \geq Al\ Al$). In the space of the possible substitutions that can exist in a given mineral (and that are in principle independent), there is a law of variation of chemical composition that is relatively simple involving various possible substitutions: it is called *paired substitutions*. Variations of the chemical composition of a given mineral appear to be an excellent marker of a geological phenomenon.

1.2 CLASSIFICATION OF THE MINERALS

There are currently around 4170 known mineral species. Among these minerals, about 50 are common rock-forming minerals. The common minerals of economic importance, forming the ores, are about 70 to 80.

Classically (*Dana's New Mineralogy*, 9th edition, 1997; Strunz, *Mineralogical Tables*, 9th edition, 2006) minerals are classified into 9 classes based on their chemical composition. According to a rather old counting of 2300 minerals, the distribution would be:

1 native elements	50
2 sulfides and sulfosalts	350
3 halides	40
4 oxides and hydroxides	220
5 carbonates [$(CO_3)^{2-}$]	100
nitrates [$(NO_3)^-$]	8
borates [$(BO_3)^{3-}$]	100

6 sulfates [$(SO_4)^{2-}$]	250
chromates [$(CrO_4)^{2-}$], molybdates [$(MoO_4)^{2-}$]	20
tungstates [$(WO_4)^{2-}$] – tellurates [$(TeO_4)^{2-}$]	13
7 phosphates [$(PO_4)^{3-}$]	
arsenates [$(AsO_4)^{3-}$]	
vanadates [$(VO_4)^{3-}$]	350
8 silicates [$(SiO_4)^{4-}$]	700
9 organic compounds	a few tens

Most of the rock-forming minerals are silicates and the number of those belonging to other classes is quite limited:

1   native elements: graphite*
2   sulfides and sulfosalts: pyrite*, pyrrhotite*
3   halides: halite, fluorite
4   oxides: spinel group, and among them magnetite* and chromite*, hematite*, ilmenite*, rutile, periclase, corundum, perovskite – and hydroxides: *goethite, limonites, gibbsite,* diaspore, *boehmite,* brucite
5   carbonates: calcite, dolomite
6   sulfates: gypsum, anhydrite, barytine
7   phosphates: apatite, monazite

(in the list above, the opaque minerals, which cannot be determined with the transmission microscope, are marked with an asterisk*, species not treated here are noted in *italics*).

Silicates are the most abundant group of minerals. They constitute over 90% of the Earth's crust. The feldspar group represents about 60% and that of silica (mainly quartz) 10 to 13%.

The fundamental structural unit of silicates is a tetrahedron $(SiO_4)^{4-}$, where the center of the tetrahedron is occupied by a silicon atom and the 4 apexes by oxygen atoms. The silicates are classified according to the arrangement of these $(SiO_4)^{4-}$ tetrahedra.

• isolated tetrahedra (**nesosilicates**, also called orthosilicates): olivine, monticellite, humite group, zircon, titanite, andalusite, kyanite, sillimanite, grenat, staurolite, chloritoid, sapphirine, topaz;
• isolated double tetraedra sharing an apical oxygen (**sorosilicates**): epidote group (pistachite, clinozoisite, zoisite, allanite, piemontite, lawsonite, pumpellyite), vesuvianite, axinite, melilite, låvenite;
• tetraedra associated in a ring (**cyclosilicates** or ring silicates); there are 3-, 4- and 6-members rings: cordierite, tourmaline, beryl, eudyalite; tetraedra linked into chains (**inosilicates** or chain silicates), there are single chains

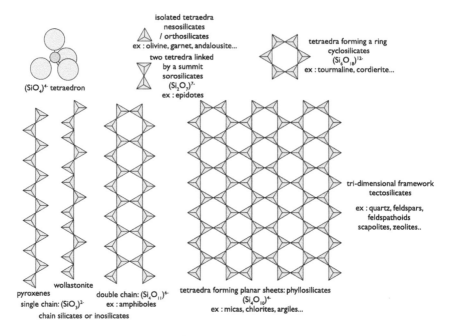

Figure 1.8 Structure of the silicates.

(pyroxenes, wollastonite, aenigmatite) and double chains: (amphiboles, astrophyllite);

•   tetraedra forming two planar sheets (**phyllosilicates** or sheet silicates): mica group, chlorite group, serpentine, talc, stilpnomelane, clay minerals, prehnite, *apophyllite*;

•   tetraedres forming a tri-dimensional framework (**tectosilicates** or framework silicates): quartz, feldspars, feldspathoids, analcime, scapolite group, *zeolite group*.

1.3   FACTORS OF OCCURRENCE OF MINERALS

The occurrence of minerals in a rock depends on physical and chemical factors.

1.3.1   Physical factors

Physical factors are conditions imposed from the outside during the formation the rock: temperature, pressure (lithostatic pressure that depends on the depth and directional pressures) and fluid pressure. These factors, pressure on the rock and fluid pressure, are more or less independents. It is widely accepted that the oriented pressures are not important in the equilibrium between minerals. They are however very important in the dynamics of growth of the minerals. The concept of **metamorphic facies** (or mineral facies) integrates these various physical factors.

The metamorphic facies are defined by one or several critical mineralogical parageneses (a paragenese is a mineral association in equilibrium). This concept was first introduced for the metamorphic rocks (Eskola, 1920, 1927). About 14 facies and sub-facies are currently recognized:

The *zeolite facies* (Coombs, 1960) includes a first sub-facies which covers the field of diagenesis and early metamorphism and is defined by the index minerals heulandite (or its polymorph clinoptinolite, these two minerals are zeolites of formula $(Ca, Na)_2 Al_2Si_7O_{18} \cdot 6H_2O$) or by the critical association analcime + quartz.

A second sub-facies is much more metamorphic: the previous associations no longer exist; it is characterized by the presence of albite $NaAlSi_3O_8$ and the association laumontite + quartz (laumontite is a zeolite of formula $CaAl_2Si_4O_{12} \cdot 4H_2O$).

The *prehnite – pumpellyite facies* is defined by the associations prehnite + quartz and pumpellyite + quartz. Jadeite and glaucophane are totally absent. Epidote may be present, and so, its appearance indicates a transition towards the greenschist facies.

The *greenschist facies* (Eskola, 1920) is characterized by the association of a non aluminous calcic-amphibole (tremolite – actinolite series) and a sodic plagioclase (An <20). The association epidote + quartz is common. Biotite may appear in the higher metamorphic part of this facies.

The *blueschist facies* (or glaucophane schist facies) (Eskola, 1929) is distinguished by the presence of a sodic amphibole, glaucophane. The presence of lawsonite and/or the association jadeite + quartz is used to define two sub-facies. Pumpellyite may exist in the less metamorphic part of this facies. The white micas are not muscovite but phengite. Biotite and feldspar are excluded.

The *white schist facies* (Chopin, 1984) is characterized by the association talc + kyanite (equivalent to magnesian chlorite at lower pressure). Coesite, high pressure polymorph of silica, is known in some parageneses, implying pressures above 25 kb.

The *albite-epidote amphibolite facies* (Becke, 1921) is characterized by a association of a hornblende (aluminous calcic amphibole), sodic plagioclase (An <20), epidote and quartz.

The *amphibolite facies* (Eskola, 1915) is characterized by the association hornblende + calcic plagioclase (An> 20). Clinopyroxene may occur but no orthopyroxene.

The *granulite facies* (Eskola, 1929) is defined by the critical association plagioclase (calcic) + orthopyroxene; garnet (of the almandine – pyrope series). Biotite and primary hornblende are present in the hornblende granulite sub-facies: The pyroxene granulites sub-facies is characterized by the complete absence of (primary) hydrous minerals.

Pyroxene-hornfels facies (Eskola, 1915) differs from the granulite facies by the absence of garnets of the almandine – pyrope family.

Sanidinites facies (Eskola, 1920) (or sanidine facies) is characterized by the presence of sodi-potassic mixed feldspar (sanidine) of very high temperature (and low pressure). The association of *mullite + sanidine + tridymite* (often altered to quartz) + cordierite + glass in metapelite is characteristic.

Eclogite facies (Eskola, 1920) is defined by the critical association of a pyrope-rich garnet and a jadeite-rich pyroxene (omphacite) (+ quartz). Plagioclases are totally absent in this facies.

The concept of metamorphic facies (the term "mineral facies" would be more appropriate) may easily be extended to igneous rocks: a granodiorite crystallizes in the amphibolite facies, charnockite in the granulite facies and a basalt in the pyroxene hornfels or sanidinites facies ...

The grade of metamorphism may sometimes by assessed in a less precise and rigorous way, by using the expressions low, medium or high grade metamorphism (Winkler, 1965–1979). This concept is actually close to older concepts of epizone, mesozone and catazone of Grubenmann (1904). In practice, in the case of metamorphism of low to medium pressure, we refer here to low grade metamorphism (or epizonal) up to the biotite isograd, of medium grade metamorphism (or mesozonal) from the isograd biotite in up to the isograd muscovite out, the limit of high grade metamorphism can be taken at the disappearance of primary muscovite.

Very generally, this concept of metamorphic facies is interpreted in terms of temperature – (lithostatic) pressure (considered equivalent to the depth) (Figure 1.9).

This diagram, very commonly used, however, contains some ambiguities. Some reactions do not involve water, others depend on water pressure. It is often assumed that the fluid pressure is essentially the water pressure and that it is equal to the lithostatic pressure. This last point is sometimes questionable and deserves at least to be discussed. On the other hand, the assumption that the fluid is essentially hydrated is certainly not proven (and probably wrong) in the granulite facies, or maybe even from the partial melting isograd.

1.3.2 Chemical factors

The chemical factors can be classified into parameters more or less linked to the fluid phase (or at least conveyed by it) and parameters directly related to the chemical composition of the rock as we it see today.

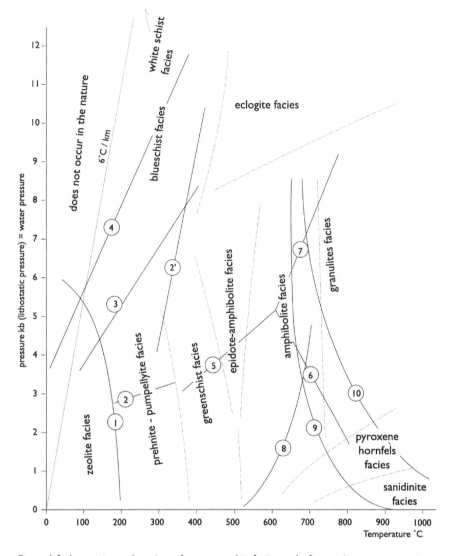

Figure 1.9 Approximate location of metamorphic facies and of some important reactions in the PT diagram.

1 analcime + quartz (low temperature) = albite + H_2O (higher temperature)

2 laumontite + 2 quartz + H_2O (low pressure) = lawsonite (high pressure)

2' upper limit of stability of lawsonite that is replaced at higher temperature by epidote

3 calcite (low pressure) = aragonite (high pressure)

4 albite (low pressure) = jadeite + quartz (high pressure)

5 andalusite (low pressure) = kyanite (high pressure)

6 andalusite (low temperature) = sillimanite (high temperature)

7 kyanite (low temperature) = sillimanite (high temperature)

8 muscovite + quartz (low temperature) = potassium feldspar + silicate of alumina + H_2O (high temperature)

9 solidus of granitic rock (water saturated)

10 solidus of an olivine tholeiite à (water saturated)

1.3.2.1   Parameters linked to the fluid phase

Parameters like fugacities (fH2O, fCO2, fS2 and fO2) belong to this; they are particularly important factors of formation and evolution of the rocks and their minerals.

For instance, the crystallization of amphibole, a hydrous mineral, or of pyroxene, anhydrous mineral, depends on water fugacity and, thus the evolution of a magmatic series is linked to the appearance of one or other of these minerals. The silica content of pyroxene is about 50–55%, that of an amphibole ranges from 40–45%. The silica content of the derived magmas will be greater or smaller depending whether one or other of these minerals appears.

Another example in a metamorphic domain is given by Buddington (1963) who describes metric alternating beds of amphibolite facies rocks (high water fugacity) and granulite facies rocks (very low water fugacity).

Oxygen fugacity controls the presence of graphite or the oxidation grade of iron or manganese, and thus the appearance of such minerals as hematite or magnetite or epidotes.

Fugacity of CO_2 controls the metamorphism of impure carbonate rocks: according to whether it is high or not, the carbonates are preserved or transformed into calcic or calcic and magnesian silicates.

Fluid inclusions in granulite facies rocks are rich in CO_2 (Touret, 1977, 1981) and sometimes are only made of CO_2 (granulites of southern India). Various models have been presented to explain the role of such fluid in the development of this facies: preferential dissolution of water in the magma or supply of CO_2 "from a deep source" ....

A high sulfur fugacity favors the crystallization of iron sulfides, pyrite and/or pyrrhotite, and thus, the iron contained in these sulfides do not participate in the formation of the silicates.

1.3.2.2   Chemical composition of the rock

The rocks are classified into five major families:

1   The **alterites and residual rocks,** which are rocks where alkalis, calcium, magnesium are more or less leached and are thus enriched in iron and aluminum. These rocks are often rich in clay, illite and kaolinite. Bauxites in particular, are made of iron and aluminum oxides and hydroxides: goethite (FeO(OH)), limonites (FeO(OH), nH_2O), gibbsite ($Al(OH)_3$), diaspore ($\alpha AlO(OH)$) and boehmite ($\gamma AlO(OH)$).

2   Sedimentary rocks **are formed by deposition and diagenesis on the surface of the Earth. They are classified into** terrigenous clastic rocks **and chemical and biochemical** rocks:

Clastic rocks may contain any mineral. In practice the nature of the minerals reworked in sedimentary rocks depends on several factors:

greater or lesser alterability, ability to endure a more or less long transportation, which is a character linked to their hardness and their brittleness (presence of cleavage) and of course of their frequency and abundance in potential source areas. Sedimentary sorting separates the fine fraction, mostly clay (mudstone, shale) from the coarser clastic fraction (mainly quartz but also feldspar, mica). An extreme example of sedimentary sorting is given by the beach or fluvial "black sands" that may be of economic interest for their deposits and that concentrate "heavy minerals", like garnet, kyanite, zircon, monazite, magnetite, chromite, ilmenite, cassiterite, gold, platinum, diamond, ruby, sapphire. Some clastic rocks contain a more or less important proportion of carbonates, either as clasts, cement or matrix.

Chemical and biochemical rocks have much more particular compositions. Chemical rocks form from direct precipitation (from seawater, brackish or freshwater): some carbonate rocks (dolostone), evaporites (gypsum, rock salt, potash), some siliceous rocks (chert, flint), (siliceous and carbonated) oolitic ironstones and manganese ores. Biochemical rocks are the result of biochemical activity of living organisms (corals, molluscs, foraminifera, stromatolites): many carbonate rocks (limestone), some siliceous rocks (radiolarite, diatomite) and carbonaceous rocks (coal, oil).

The size of the grain of sedimentary rocks is often too small for their mineralogy to be studied under the petrographic microscope. Their study then requires X-ray diffraction.

3 **Magmatic or igneous rocks** are the result of the cooling and solidification (most times crystallization) of a magma (most often a silicate magma). They are classified according to their level of emplacement in the Earth's crust: extrusive or *volcanic rocks, hypabyssal rocks* and intrusive or *plutonic rocks*. A second classification is done by the relative proportion of dark- (ferromagnesian minerals, oxides) to light colored (quartz, feldspar, feldspathoids) minerals, expressed as color index. Ultramafic rocks are rocks containing more than 90% of colored minerals.

Chemical terms used to designate the igneous rocks are:

- *acidic rocks* with a SiO_2 content higher than 66 wt%); such rocks are also riche in alkalis Na, K, and often in iron (sometimes ferric iron) and hygromagmaphiles elements (F, B, Zr, Nb, Sn, W); acid rocks are often felsic rocks, which is a different notion referring to their mineralogy and not to their chemistry;
- *intermediate rocks,* with a SiO_2 content comprised between 52 and 66 wt%;
- *basic rocks* (SiO_2 between 45 and 52 wt%);
- *ultrabasic rocks* ($SiO_2 < 45\%$).

The notions of ultramafic and ultrabasic rocks are distinct: an ultramafic rock is not necessarily ultrabasic and vice versa.

The rocks of the last two groups are rich in Ca, Fe, Mg, Ti. Their minerals will be more or less calcic plagioclase, amphibole, pyroxene, olivine. Let us recall the importance of water in the crystallization of amphibole or pyroxene.

1.3.2.3 Silica saturation of igneous rocks

After oxygen, silicon is the most abundant element in most igneous rocks – which are mainly formed of silicates. Minerals of igneous rocks fall into two groups:

- silica saturated minerals that may be in equilibrium with quartz;
- silica undersaturated minerals, which react with quartz to form a silica saturated. Among the latter, feldspathoids (feldspathoid + quartz = feldspar) and magnesian olivine (olivine + quartz = orthopyroxene) are particularly important Another example is the perovskite (perovskite + quartz = titanite).

So the rocks may be classified according to this the silica saturation:

- quartz present in an amount greater than 5%: silica over-saturated rocks;
- less than 5% of quartz or feldspathoid: silica-saturated rocks;
- feldspathoid present in an amount greater than 5%: silica under-saturated rocks.

1.3.2.4 *Alumina saturation of igneous rocks*

After oxygen and silicon, aluminum is the second most important element in most igneous rocks. Feldspars and feldspathoids are most often the major aluminum bearing minerals. Given the composition of feldspar ($KAlSi_3O_8$, $NaAlSi_3O_8$, $CaAl_2Si_2O_8$) and feldspathoids ($KAlSi_2O_6$, $NaAlSiO_4$), there may be relative excess or deficit of aluminum to alkalis and calcium; this relative excess of aluminum or alkalis will be expressed by specific minerals:

Al > 2Ca + Na + K: *peraluminous rocks*; this excess of aluminum is reflected by the presence of aluminous minerals other than feldspars: muscovite, garnet, cordierite, alumina silicates ...;

2Ca + Na + K > Al > Na + K: *meta-aluminous* rocks; such rocks contain feldspar, possibly biotite and hornblende, but no peraluminous minerals nor alkali mineral;

Al < Na + K: *(per)alkaline rocks* contain alkali ferromagnesian minerals (sodic amphibole and/or pyroxene, sometimes sodic zircono-silicates and: or titano-silicates). Most of the alkaline rocks are sodic. There are also potassic (per-) alkaline rocks. Particular mention should be made for potassic rocks rich in mafic minerals (kimberlites, some lamprophyres). Rarer are the rocks that show a relative deficit of aluminum to calcium: wollastonite ($CaSiO_3$) is a mineral sub-saturated in aluminum compared to anorthite ($CaAl_2Si_2O_8$); it is found in some rare alkali granites (Corsica) or some nepheline syenites (Alnö). Peralkaline rocks only represent 1 to 2% of the igneous rocks.

The concepts of saturation in silica and in alumina are independent: some granites (silica-saturated rocks) are peralkaline (and contain Na-amphibole); nepheline syenite (rock under-saturated in silica) are saturated in alumina (miaskitic syenites); other nepheline syenites are under-saturated in both silica and alumina (agpaitic syenites).

These notions of silica and alumina saturation were originally defined for the igneous rocks. They can be extended to the other rocks.

4   **Metamorphic rocks** result from the endogenous transformation of other rocks: sedimentary rocks (para-derived metamorphic rocks), igneous rocks (ortho-derived metamorphic rocks) or other metamorphic rocks. As metamorphism is nothing but a transformation due to changes of physical conditions, without any change in chemical composition other than the content of volatile elements, the mineralogy of metamorphic rocks reflects the chemical composition of rocks from which they derive.

Among the clastic sedimentary rocks, the pelitic rocks are often rich in clays, and thus peraluminous. Metamorphism develops in such rocks aluminous minerals: alumina silicates, muscovite, garnet, cordierite, staurolite, chloritoid .... The more or less impure sandstones are generally much less rich in alumina, while remaining mostly peraluminous: the latter aluminous minerals appear less frequently or, at least, in smaller quantities.

The rocks that initially contained carbonate (calcite and/or dolomite), present a varied mineralogy depending on the relative proportion of the initial carbonates and the other minerals, that may contain alumina (clay, feldspar) or not (quartz).

• if the proportion of carbonate is low, the rock will be made of quartz, feldspar, mica, aluminous silicates; calcium is only expressed in garnet or plagioclase by a by a more or less large proportion of grossular molecule or anorthite;
• a higher proportion of calcium is expressed by specific mineral, calcic and calcic and magnesian silicates (amphibole, diopside, grossular-richer garnet, anorthite-richer plagioclase): such metamorphic rocks are calc-silicates-gneiss (or calc-silicate rocks). In this family of rocks, one can distinguish members richer in aluminous minerals (calcic plagioclase, epidote), which contained before metamorphism more or less clay minerals (such rocks were originally marls) and terms poorer in aluminous minerals, which were originally calcareous sandstone ("calc-silicate-quartzites"). In such rocks there is no free primary carbonates – or they are in a very small proportion;

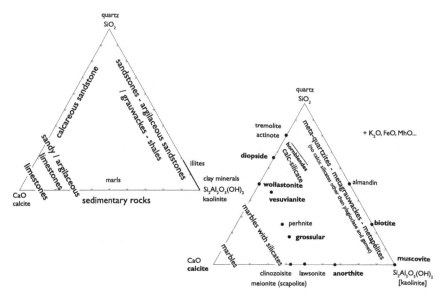

Figure 1.10 Carbonate rock.

The minerals that are stable in medium to high grade metamorphism are in bold; those stable only at low grade metamorphism are in normal font.

- The primary free carbonate will become abundant when the initial proportion of carbonates was larger: Such rocks abundant carbonates are
  marbles. The presence of a small proportion of clay, quartz and other
  minerals in the initial rock is reflected by the presence of calcic silicates,
  which remain minors in front of the carbonates. This carbonate is usually calcite. Dolomite, if it was present initially, often reacts, rather than
  calcite, with silica and silicates to form magnesium silicates like talc,
  tremolite, phlogopite, diopside, forsterite and/or spinel) + calcite.
- Metamorphism transforms a rock originally composed almost only of
  carbonate, with a very small proportion of silica clays, into a nearly
  pure marble.

5   **Hydrothermal rocks** are endogenous rocks formed through the circulation of a (usually hydrated) fluid: either by direct precipitation (quartz
   veins, mineral veins, alpine veins with axinite, epidote, albite) or by
   transformation by these fluids of preexisting rocks. The metasomatic
   process, with introduction and leaching of elements, leads to the formation of specific rocks that are often almost monomineralic: secondary
   dolomites, chloritites, albitites ...

Greisens are mainly formed of quartz + white mica, result from the
leaching of alkali feldspar and are developed mostly at the expense of acid

plutonic rocks. This leaching of the alkalis is accompanied by an introduction of elements like tin, tungsten, lithium and particularly volatile elements like boron (tourmaline) and fluorine (topaz, fluorite).

Skarns are rocks mainly composed of calcic silicates that are developed at the contact between two chemically incompatible media, most often between calcic rocks (often carbonate rocks but also basic or ultrabasic rocks) and a siliceous rocks (igneous rocks often of the family of granites). Skarns contain the same minerals as calcic carbonate rocks (calcic garnets, pyroxenes, amphiboles), some minerals like wollastonite and scapolite that are not exclusive of skarns, are much more common in the skarns; there are also minerals that are specific to the skarns, like silica under-saturated calcic minerals.

1.4 PLAN ADOPTED IN THIS GUIDE

Most books of systematic mineralogy present minerals in the order of the classical classifications by Dana (1997) or by Strunz (2006) which are based on the chemical composition and structure. The silicates are classified by the arrangement of $(SiO_4)^{4-}$.tetrahedra.

The rock-forming minerals, which are mostly silicates, are presented here by chemical affinity, which allows us to regroup minerals that have similar occurrences and that are frequently associated.

1   major framework silicates: quartz, feldspaths, feldspathoids, ubiquitous minerals which are the main bearers of aluminum and alkalis;
2   major ferro-magnesian minerals: micas, chlorites, amphiboles, pyroxenes, olivines;
3   aluminous minerals: minerals resulting from the metamorphism rocks of clay-rich sedimentary (pelites, shales) and peraluminous igneous rocks;
4   calcic, magnesian and calc-magnesian minerals: minerals of more or less impure metamorphic marbles, calc-silicate-gneisses; skarns and some calcium-rich igneous rocks;
5   accessory minerals which are the bearers of minor elements like B, P, Ti, Zr ...;
6   some minerals more common in sedimentary rocks; only those which are recognizable under the microscope will be here discussed;
7   minerals of the ore deposits are mostly opaque minerals whose study requires the use of the reflected light microscope. Some minerals recognizable with the petrographic microscope are presented.

Chapter 2

Observations with the petrographic microscope

Rocks and minerals are observed as parallel-sided thin sections through the petrographic microscope, a transmit light microscope using polarized light. By convention, the thickness of a thin section is 30 micrometers (μm). The thin section may be thicker for some specific uses, such as the study of fluid inclusions. The opaque minerals are studied using the reflected light microscope. The use of these two types of microscopes requires specific training and experience.

The purpose of such observations is to identify the minerals forming the rock, study the relationships between these minerals and the structure of the rock and thus reconstruct its own history.

The first observation of the phenomenon of double refraction was due to Bartholin around 1698 (Touret, 2006). But it was the discovery of the polarization of light by Malus (1807) that provided to this author an explanation of this phenomenon (1808). In the early nineteenth century Brewster (1781–1868), studying fragments of minerals under the microscope, identified the position of the optical axes, defined uniaxial and biaxial minerals, observed for the first time fluid inclusions and was able to measure the indices (of the minerals and the fluid inclusions) with an accuracy better than 0.05. The first thin sections date from Nicol (1815) and Oschatz (1852). But it is mainly due to the work of Sorby (*On the Microscopical Structure of Crystals*, 1858, first observation of fluid inclusions in thin section) and the development, by Frankenheim, of the petrograpic microscope in 1860 that this device actually entered into the methods of petrography. The late nineteenth century saw the development of the petrographic microscope (Ernst Karl Abbe, Camille Sebastien Nachet, Emmanuel Bertrand) with the technological development of the microscope and the systematic description of minerals in thin section. The German school was first prominent: Zirkel (*Microscopische Petrography*, 1876), Vogelsang (*Die Kristalliten*, 1875), Rosenbuch (*Microscopische der Mineralien und Gastein Physiography*, 1873–1877), Becke (*Becke line*, 1890). Russian Fedorov perfected the universal stage (1883). The end of the century is dominated by the French School (Ecole des Mines, Museum d'Histoire Naturelle, College de France): theory of crystalline optics and its application to petrographic microscope was firmly established by the work of Fouque and Mallard (Cristallographie géométrique et physique (*Geometrical and Physical Crystallography*), 1884–1885), mineral synthesises by Daubree at the Ecole des Mines, works: of Des Cloizeaux (Méthode de détermination des plagioclases (*Method for determining Plagioclases*), 1875), Fouqué (Fouqué and Michel-Lévy: Minéralogie micrographique (*Micrographic Mineralogy*), 1879), Michel-Lévy (Etude sur la détermination des feldspaths dans les plaques minces au point de vue de la classification des roches (*Study on the determination of feldspars in thin sections from the point of view of the classification of rocks*), 1894–1904) and Lacroix (Michel-Lévy and Lacroix: les Minéraux des roches (*The Minerals of the Rocks*), 1888). The main methods of microscopy applied to the study of minerals and rocks were acquired in 1900.

2.1   INDICATRIX (REFRACTIVE INDEX ELLIPSOID)

Light is an electromagnetic vibration. In an optically isotropic medium, the electromagnetic wave propagates in a straight line and the electromagnetic vibration is perpendicular to the direction of propagation. The velocity of propagation of the light in vacuum is c. The velocity of propagation of the light in a given medium is c/n, where n is the index of this medium. In an optically isotropic medium, light propagates without distortion, with the same velocity in all directions.

In an isotropic medium, the electromagnetic vibration may be the same in all directions, it is called *non-polarized light* or *circular polarized light*. In the case of *plane polarization*, the electric vector (and the magnetic vector that is perpendicular to it) vibrates in a single plane. In the case of *elliptical polarization*, the electric and magnetic vector rotate around their axis and their amplitude varies so that it describes an ellipse. Plane polarized light may be obtained by different processes: an appropriately cut prism of Iceland spar (transparent crystallized calcite) called Nicol prism or Nicol (1829) has been used; and nowadays we use polaroids made of organic crystals oriented included in a transparent film (1929–1938).

In an anisotropic medium, the light vibration does not propagate without distortion. *But Fresnel theorems show that for any mineral section, there are two perpendicular directions in which light propagates without distortion at speeds $c/n\gamma$ and $c/n\alpha$.*

If we change this section in all directions, the locus of the vectors $n\gamma$ and $n\alpha$, is an ellipsoid, called the **refractive index ellipsoid** or **indicatrix** (Figure 2.1).

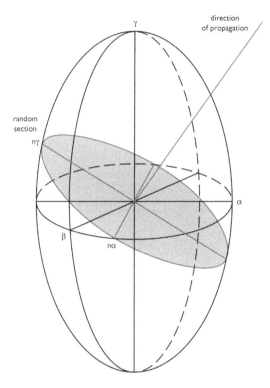

Figure 2.1 Indicatrix.

This ellipsoid is defined by its three principal axes α, β and γ for biaxial minerals (Notation by Deer, Howie and Zussman, 1966–1992). Those indexes may also be noted respectively Nα, Nβ and Nγ, or X, Y, Z in Tröger (1971)) or Np, Nm and Ng (French notation). The notation of the indexes for an uniaxial mineral are ε (principal axis of symmetry) and ω (perpendicular to it) (note that this anglo-saxon notation is rather inconsistent).

Birefringence is the quantity $\Delta = \gamma - \alpha$ for biaxial minerals or $|\varepsilon - \omega|$ (absolute value) for uniaxial minerals, (or, in any case, Ng – Np according to the French notation).

The indicatrix has the symmetry of the medium and its axes of symmetry coincide with those of the mineral:

• In a cubic mineral, it is a sphere: a cubic crystal is optically isotropic.
• In a mineral having an axis of symmetry of order greater than 2 (hexagonal, trigonal, tetragonal systems), the indicatrix is an ellipsoid of revolution.
• In the orthorhombic, monoclinic and triclinic systems, the indicatrix is a triaxial ellipsoid.

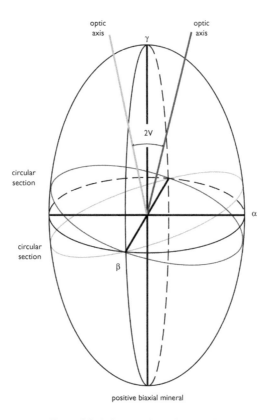

Figure 2.2 Indicatrix: biaxial mineral.

If the indicatrix is a triaxial ellipsoid (orthorhombic, monoclinic and triclinic systems), there are two *circular sections*. These sections are optically isotropic. Directions perpendicular to the two circular sections are the **optic axes**: the mineral is **biaxial**.

The acute angle between the optic axes is the **optic angle** or **2V** angle.

If γ is the bisector of the acute angle of the optic axes, the mineral is **positive biaxial**; α is then the bisector of the obtuse angle of the optic axes.

If α is the bisector of the acute angle of the optic axes, the mineral is **negative biaxial**.

Birefringence of any section (Figure 2.1) is:

$$n\gamma - n\alpha = (\gamma - \alpha) \sin\theta \sin\theta'$$

where θ and θ' are the angles between the normal to this section and the two optical axes.

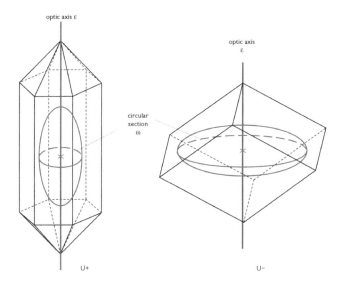

Figure 2.3 Indicatrix: uniaxial minerals.

In the hexagonal, trigonal, tetragonal systems, the indicatrix is an ellipsoid of revolution: there is only one optic axis which is the axis of revolution, the mineral is called **uniaxial**. A section perpendicular to the optic axis (axis of revolution) is optically isotropic. Index along the optic axis is called ε, index along the circular section is called ω.

If $\varepsilon > \omega$, the indicatrix is an elongated (prolate) ellipsoid («cigar-shaped ellipoid»), the mineral is called **positive uniaxial** (for example: quartz).

If $\varepsilon < \omega$, the indicatrix is an flattened (oblate) ellipsoid («pancake- or disk-shaped ellipsoid»), the mineral is called **negative uniaxial** (for example: calcite, tourmaline).

2.2 THE PETROGRAPHIC MICROSCOPE

Basically, a polarizing microscope in transmit light includes:

*   an **illuminator,** source of a parallel beam of white non-polarized (circular polarized) light;
*   the **lower polarizer** (or Nicol) that transforms the incident beam into plane polarized light. This first polarizer may be rotated. The cross hairs of the reticle (at right angle) materialize the polarization plane of the polarizers. Depending on the construction of the microscope, polarization plane of the first polarizer may be E–W or N–S;

- a **stage** bearing the thin section. The stage can rotate freely around the axis of the microscope and can carry a mobile holder (mechanical stage) for moving the thin section and precisely locate a point. The edge of the stage is graduated so that the angles can be measured; most models are also equipped with a vernier for precise measures;
- the **objective**, actually there are a series of objectives with different magnifications placed on rotating nosepiece. The observations are made from the lowest magnification to the highest one. You must take care to handle the nosepiece at the base and not by objectives, which may decenter the objectives;
- the **upper polarizer,** or **analyser,** is removable and whose polarization plane of is at right angle to that of the lower polarizer. You verify that the two polarizers are at right angles by observing that no light emerges from the analyzer when the two polarizers are crossed;
- the **ocular.**

Various removable accessories are installed on the microscope:

1 a diaphragm placed over the light source used to modulate the size of the light beam; it is used for increasing the contrast between two fields of the image (for instance for a Becke line);
2 removable auxiliary condenser lenses which converts the incident parallel light beam (orthoscopic illumination) into a conical one (conoscopic illumination); it is used to make interference figures in convergent light;
3 accessory (compensator) plates, that are introduced into a slot located between the objective and the analyzer. The accessory plates have a known birefringence. The most useful is a quartz plate whose birefringence is equal to $\delta = e(n\gamma - n\alpha) = 0,018$ (for a thickness of 30 micrometers) (which corresponds to purple/indigo color that marks the boundary between the red at end of the first order and the blue of the beginning of the second order in Michael Lévy's interference color chart). The accessory plate is used to determine the sign of elongation of a section or the optical sign of a mineral;
4 the Bertrand lens, removable, is used to observe interference figures in convergent light.

An *objective* lens must be *centered* so that the axis of the microscope coincides with the rotation axis of the stage. It is centered either through rotating rings located on the objective or small screws at the lens collar or in the rotating nosepiece. The method involves: identify a distinctive point in the thin section, place it at the intersection of the cross hairs of the reticle and rotate the stage of 180°. If the objective is not centered, the identified point is no more located at the intersection of the cross hairs. Then move the

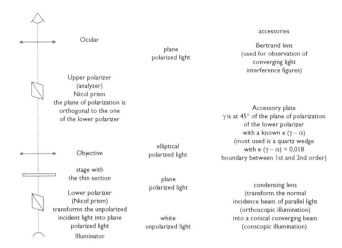

Figure 2.4 The petrographic microscope.

objective so that the intersection of the cross hair moves halfway between its initial position of the identified point and its position after rotation, and do it again as many times as necessary. The exercise may be especially irritating.

Generally, a rock is made of several minerals. A thin section shows different sections of each mineral. The most difficult task for the beginner, is to recognize the different sections of the same mineral: for instance, a muscovite cut parallel to the cleavage does not look the same as muscovite cut perpendicularly to the cleavage. **The observations intend to reconstruct, from these different sections, a 3D image of the mineral, its structure (forms, cleavages, twins), its indicatrix and the position of the indicatrix in relation to its structural elements.**

For this purpose, there are three types of observations:

1   Observations with one Nicol, the lower polarizer, the incident beam is plane polarized, these observations are called **observations in parallel polarized light** (here abbreviated into **PPL**) (it is called in French "observation in natural light" or NL).
2   Observations with two Nicols, the lower and upper (analyser) polarizers crossed at right angle. In the absence of a thin section, no light should be observed through the ocular. When a thin section is interposed between the lower and upper polarizers, the vibration coming out of the analyser is not nul. The theory of this phenomenon will be explained in the following paragraph. These observations are called **observations in cross polarized light** (here abbreviated into **CPL**) (it is called in French "observation in polarized light" or LP).

3   Addition of the condenser converts the incident parallel beam into a cone beam and creates an interference figure that can be observed by either removing the ocular or by adding the Bertrand lens. There is no particular name for such observation in English; in French it is called **"observations in convergent light"** (LC).

Although opaque minerals are not determined with the transmit light microscope we may still have some indication in considering the form of opaque minerals (cubic habit for magnetite and pyrite, elongated habits for ilmenite, for instance) and also by removing the lamp from its slot and illuminating the thin section from above: One recognizes the yellow brass color of pyrite, bright yellow to orange of chalcopyrite, the grays of ilmenite and magnetite. This does not replace a serious determination with the reflected light microscope.

2.3   CRYSTALLINE PLATE WITH PARALLEL FACES IN CROSSED POLARIZED LIGHT

Observations under the petrographic microscope are interpreted by the interaction between a plane polarized light wave and a plate with parallel faces of a given mineral.

Let us consider a plate with parallel faces characterized by the perpendicular directions in which the light propagates without distortion at velocities $c/n\gamma$ and $c/n\alpha$. These two directions we will be used as coordinate axes Ox and Oy.

The incident vibration, after the (lower) polarizer, is plane polarized:

$$OP = a \sin 2\pi \, t/T$$

where a is the amplitude of the vibration and T its period.

This vibration makes an angle α with the Ox axis.

If we decompose this vibration on the two axes Ox and Oy

$$Ox = a \cos \alpha \sin 2\pi \, t/T$$
$$Oy = a \sin \alpha \sin 2\pi \, t/T$$

This vibration passes through the plate of e thickness with a velocity $c/n\gamma$ along Ox and velocity $c/n\alpha$ along Oy. The duration of the crossing will be $(e \, n\gamma/c)$ along Ox and $(e \, n\alpha/c)$ along Oy.

At the exit of the plate, the equation of the vibration will be:

$$Ox = a \cos \alpha \sin 2\pi \, (t - e \, n\gamma/c)/T$$
$$Oy = a \sin \alpha \sin 2\pi \, (t - e \, n\alpha/c)/T$$

The vibration that propagates along Ox is retarded compared to the one that propagates along Oy of a quantity:

$$2\pi\ \phi = 2\pi\ e\ (n\gamma - n\alpha)/cT = 2\pi\ e\ (n\gamma - n\alpha)/\lambda$$

where $\lambda = cT$ is the wave length.

The two vibrations are recomposed at the exit of the plate into an elliptically polarized vibration: *a plane polarized wave is transformed by the passage through a parallel plate of an anisotropic medium into an elliptically polarized wave.*

Means a change of variables, the equation of this wave at the exit of the plate is:

$$Ox = a\ \cos\ \alpha\ \sin2\pi\ (t/T - \phi)$$
$$Oy = a\ \sin\ \alpha\ \sin2\pi\ t/T$$

with $\phi = e\ (n\gamma - n\alpha)/\lambda$

This vibration then passes through the analyser (upper polarizer) whose polarization plane makes an angle β with the one of the (lower) polarizer. The vibration in the plane of the analyser will be:

$$OA = Ox\ \cos\ \beta + Oy\ \sin\ \beta = a\ \cos\ \alpha\ \cos\ \beta\ \sin2\pi\ (t/T - \phi)$$
$$+ a\ \sin\ \alpha\ \sin\ \beta\ \sin2\pi\ t/T$$

Practically the Nicols are at right angle: $\beta = \alpha + \pi/2$

$$OA = a\ \cos\ \alpha\ \cos\ (\alpha + \pi/2)\ \sin2\pi\ (t/T - \phi) + a\ \sin\ \alpha\ \sin\ (\alpha + \pi/2)\ \sin2\pi\ t/T$$
$$OA = a\ \cos\ \alpha\ \sin\ \alpha\ \sin2\pi\ (t/T - \phi) + a\ \sin\ \alpha\ \cos\ \alpha\ \sin2\pi\ t/T$$
$$OA = a\ \sin\ \alpha\ \cos\ \alpha\ [\sin2\pi\ t/T - \sin2\pi\ (t/T - \phi)]$$
$$AS\ \sin\ p - \sin\ q = 2\cos\ (p + q)/2\ \sin\ (p - q)/2$$

$$OA = a\ 2\ \sin\ \alpha\ \cos\ \alpha\ [\cos\ 2\pi\ (t/T - \phi/2)\ \sin\pi\phi]$$
$$OA = a\ \sin\ 2\alpha\ \sin\pi\phi\ [\cos\ 2\pi\ (t/T - \phi/2)]$$

The amplitude of this vibration is thus:

$$a\ \sin\ 2\alpha\ \sin\pi\phi = a\ \sin2\alpha\ \sin(\pi\ e\ (n\gamma - n\alpha)/\lambda)$$

As the intensity is proportional to the square of the amplitude, the intensity at the exit will be:

$$a^2\ \sin^2 2\alpha\ \sin^2(\pi\ e\ (\mathbf{n\gamma - n\alpha})/\mathbf{\lambda})$$

$\delta = (n\gamma - n\alpha)$ is the birefringence of this section

$$a^2\ \sin^2 2\alpha\ \sin^2(\pi\ e\delta/\lambda)$$

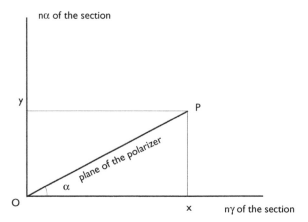

Figure 2.5 Plate with parallel faces: wave after passing through the polarizer.

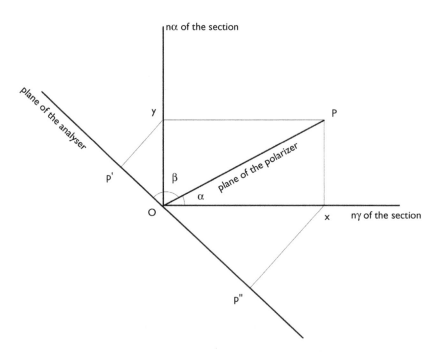

Figure 2.6 Plate with parallel faces: recomposed wave after passing through the analyser.

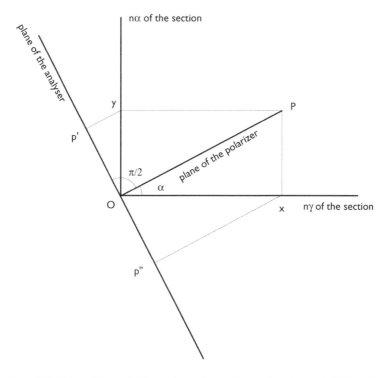

Figure 2.7 Plate with parallel faces: (lower) polariser and analyser at right angle.

or by posing $d = e\delta = e(n\gamma - n\alpha)$

$a^2 \sin^2 2\alpha \sin^2(\pi \, d/\lambda)$

The intensity of the emerging beams depends on the wavelength λ, that is to say the color. The incident light is white light; we must thus integrate the intensity over the range of wavelengths:

$I = \int_\lambda I_\lambda$

For each wavelength, the curve I_λ is a sinusoid all the more spread that λ is larger. The observed coloration is the sum of the colors obtained for each λ at $d = e \, (n\gamma - n\alpha)$ constant. In function of d, we obtain a wide range of colors known as the **Michel-Lévy interference color chart**:

The chart is divided into several *orders* by boundaries corresponding to values of $\delta = e(n\gamma - n\alpha)$ equal to a multiple of 0,018 (for a thickness of 30 micrometers). At such thickness, colors corresponding to these boundaries are purple/

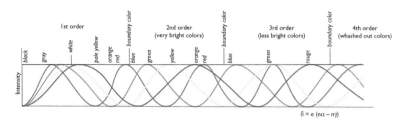

Figure 2.8 Michel-Lévy interference color chart.

indigo intermediate between red and blue (in French these colors are called "teintes sensibles" = sensitive hues). In practice, one can easily distinguish:

1   colors of the beginning of the first order going from black to various grays, more and more light up to white;
2   shades of the end of the first order: white, pale yellow, yellow, orange, red;
3   bright colors of the 2nd and 3rd order: blue, green, yellow, orange and red hues of the 3rd order are a little less vivid than the colors of the 2nd order, but the difference is not very obvious;
4   washed out colors orders from higher orders: the iridescent beige, pinkish, greenish.

Warning! When a mineral has a strong color, the interference colors are superimposed on the mineral's own colors and thus may not be very recognizable!

Variations with α

With the rotation of the microscope stage, four extinctions per turn are observed, the extinction occurs for $\alpha = k\,\pi/2$. Extinction occurs when $n\gamma$ and $n\alpha$ of the section coincide with the planes of polarization of the polarizer and analyzer (materialized by the cross hairs of the reticle). The maximum intensity is observed at 45° of the directions of extinction.

Variations with e

The thin sections are cut to a standard thickness of 30 micrometers. If we know the value of $(n\gamma - n\alpha)$ of the section of a given mineral, we can use this data to measure the thickness e of the thin section. This property is used by the technician to find out if the thin section is at the right thickness: at 30 micrometers: the interference color for quartz must be white (between pale yellow and gray of the first order).

Variations with (nγ – nα)

This value is characteristic of the section of the studied mineral. This value varies for different sections of the same mineral, as $n\gamma - n\alpha = (\gamma - \alpha) \sin\theta \sin\theta'$ (see paragraph 2.1) from 0 for the circular section(s) to a maximum value:

$$\Delta = \gamma - \alpha$$

The observed interference colors vary from black (no light) for circular sections up to a maximum value corresponding to the birefringence of the mineral: this value is characteristic of the mineral. The Michel-Lévy chart shows these interference colors as a function of the birefringence of the mineral, Δ and thickness of the thin section, e. Remember that in most cases this thickness is 30 micrometrers.

2.4   OBSERVATIONS IN PARALLEL POLARIZED LIGHT (PPL)

2.4.1   Forms, fractures, cleavages

A mineral may be *euhedral* (idiomorphic) with its own crystalline forms, or *anhedral* (without any crystalline forms).

A *cleavage* is a separation plane, more or less perfect, regular, repeated, that corresponds to a zone of weakness in the lattice of the mineral. Cleavage may be more or less good: perfect thin lamellar cleavage in muscovite, good in amphibole or in pyroxenes, poor in nepheline. The same mineral can have several cleavages, which are not necessarily the same quality: some perfect, others not. The cleavages follow the same laws of symmetry as the crystal.

A *fracture* is a break that is not planar in general, and overall is not repeated regularly, or at least, is repeated a few times. The fractures of some minerals, such as tourmaline, may be characteristic.

2.4.2   Index/refringence

The index of a mineral can be appreciated by the effect of relief of this mineral compared with the surrounding minerals. One can estimate this index by comparison to that of Canada balsam (1.535–1.542), or araldite (1.545), commonly used to glue the cover slip on the preparation, or in comparison with a common mineral like quartz (1.544–1.553).

Practically various indexes may be distinguished:

- very weak like fluorite (1,433–1,435) which shows a strong negative relief;

- weak, clearly lower than the one of quartz, like potassium feldspars (1,514–1,526);
- of the same order as quartz, like cordierite or oligoclase (plagioclase);
- moderate to strong like muscovite (1,552–1,716), amphiboles (1,599–1,730) or tourmaline (1,610–1,675);
- strong like orthopyroxenes (1,650–1,788) or kyanite (1,712–1,734);
- very strong like garnet (1,714–1834), epidotes (1,670–1,797) or zircon (1,923–2,015).

Precise estimation of the relative relief of two minerals is made by the *Becke line* method. The microscope is focused on the contact of the two minerals, using a suitable magnification and possibly stopping down the aperture of the diaphragm. If the objective is **moved away** from the thin section (*increasing* distance) a bright fringe appears and moves towards the mineral with **the higher index** (=more refracting mineral = with a higher relief). If the objective is **moved closer** to the thin section (*decreasing* distance) the bright fringe moves towards **the least refracting** mineral (=with a lower relief).

When a mineral has a high birefringence, that is to say two very different indices γ and α (ε and ω), the relief of the mineral will be very different depending on its relative position to the plane of polarization of incident light: it is the phenomenon of *change of relief* (in French: «pléochroïsme de relief»). A classical example is calcite: γ is much lower than quartz (1,486) and α much higher (1,658) (Δ = 0,172); by rotating the stage, the calcite appears either with a negative relief in comparison with quartz, or in positive relief.

2.4.3   Color – Pleochroism

When a beam of white light passes through a given mineral, part of the incident light is absorbed. The absorption coefficient depends on the wavelength. The mineral appears with the color complementary to that which has been absorbed. Some minerals are *opaque*, others entirely transparent for all the wavelengths (so they appear *colorless*), others *colored* (with a distinctive color).

In anisotropic media, the absorption coefficient depends on the direction of the light transmitted: as the incident wave is plane polarized, the mineral appears colorful, more or less colored or even shows very different colors depending on the relative position of section and of the plane of polarization – this phenomenon is **pleochroism**. For instance, the color of biotite varies from reddish brown to pale brown, that of piemontite from carmine to orange.

The most intense color is observed either in the NS direction or in the EW direction of the cross hairs of the reticle, depending on the mineral and position of the plane of polarization of the polarizer. Some minerals are the most intense colors in the same direction as the most intense shades observed in biotite: it is called *direct pleochroism*. Others, such as tourmaline, have less intense shades in the directions where those of biotite are the most intense (and vice versa): this is called *reverse pleochroism*.

2.4.4   Inclusions

Some inclusions are characteristic of the host mineral. For instance the inclusions of zircon in the cordierite are surrounded by a characteristic yellow pleochroic halo.

Others allow us to reconstitute the history of the mineral. For instance, euhedral oriented inclusions of plagioclase in phenocrysts of orthoclase witness the magmatic evolution.

Fluid inclusions are particularly interesting in petrology. Their study requires special skill and equipment.

2.4.5   Alterations

Alterations allow us to reconstruct the late history of a mineral, and thus the rock.

Some alterations are characteristic of some minerals and are thus a criterion of determination: for example cordierite is altered to yellowish isotropic chlorites; red clouding is characteristic of orthoclase.

2.5   OBSERVATIONS IN CROSS POLARIZED LIGHT (CPL)

2.5.1   Interference colors – Birefringence

The thickness of the thin sections being fixed at 30 μ, the interference colors of a mineral enable us to estimate its birefringence. The birefringence of any section ranges from 0 to $\Delta = \gamma - \alpha/(\varepsilon - \omega)$. Only the colors corresponding to the maximum birefringence $\Delta = \gamma - \alpha$ are significant. Sections perpendicular to an optical axis ($n\gamma - n\alpha = 0$) are always extinct. The sections of low birefringence, which are almost perpendicular to an optical axis ($n\gamma - n\alpha$, very low), can be confusing.

2.5.2   Anomalous interference colors

The position of the indicatrix of some minerals (chlorite, epidote, chloritoid, melilite), depends on the wavelength (in French such minerals are called "minéraux dispersifs", there is no correspondent expression in English). The above calculation integrating the light intensity as a function of λ no longer applies and the interference colors of such minerals no longer belong to the Michel-Lévy chart; such colors are referred to as anomalous colors: brown gray, olive green, Prussian blue, lemon yellow, etc. There are not many minerals that produce such anomalous colors; recognizing these anomalous colors is an important guide to their identification.

2.5.3 Position of the indicatrix – Angle of extinction

According to the above model, extinction of the studied section occurs four times per revolution of the stage, every time the nγ and nα of the section coincide with the polarization planes of the nicols. Identify these directions of extinction in relation to outstanding crystallographic elements (elongation, cleavage, twins) therefore allows us to locate the position the indicatrix in relation to the lattice; it is all the more accurate as several sections have been examined. Extinction can be parallel to one of these structural elements (parallel extinction) or occurs at some angle to them (inclined or oblique extinction). The angle between the direction of extinction (and, more specifically between γ, β and α) and these structural elements is characteristic of a given mineral and/or varies regularly depending on its chemical composition. Charts were established for accurate determinations and are commonly used for the determination of plagioclase. The precise construction of these structural elements and the indicatrix requires Fedorov universal stage.

2.5.4 Sign of elongation

The directions of extinction allow the identification nα and nγ of a section. The problem is how to recognize which one is nα or nγ. Let us consider a remarkable direction of a given mineral: elongation, cleavage, twin, etc. It is said that this direction has a *positive elongation if this direction is nγ of the section* (or at least form an acute angle small enough with nγ), *the elongation is said to be negative, if it is nα.*

To determine the sign of elongation, the stage is rotated at 45 degrees to the direction of extinction in order to have a maximum intensity. An auxiliary plate, with a known birefringence and a known γ direction is added (in the slot of the microscope) between the objective and the analyzer in SE–NW direction; γ direction of the plate has thus a SW–NE direction. The most useful is a quartz plate whose birefringence is equal to $\delta = e(n\gamma - n\alpha) = 0{,}018$ (for a thickness of 30 micrometers) which corresponds to the boundary between first and second order in the interference color chart (such quartz plate is called in French "quartz teinte sensible").

If nγ of the mineral has the same direction than the one (γ) of the accessory plate, that is to say that this is a positive elongation, retardations induced by the mineral and the accessory plate add: interference colors of

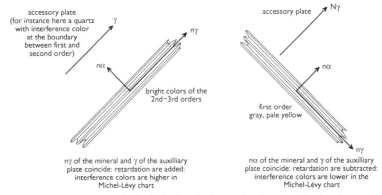

accessory plate
(for instance here a quartz
with interference color
at the boundary
between first and
second order)

γ

nγ

nα

bright colors of the
2nd–3rd orders

accessory plate   Nγ

nα

first order
gray, pale yellow

nγ

nγ of the mineral and γ of the auxilliary
plate coincide: retardation are added:
interference colors are higher in
Michel-Lévy chart

nα of the mineral and γ of the auxilliary
plate coincide: retardation are subtracted:
interference colors are lower in the
Michel-Lévy chart

nγ coincides with the elongation (for example a cleavage): positive elongation

Figure 2.9 Sign of elongation.

the mineral are added to the ones of the accessory plate; the result is that the interferences colors of the whole are higher in the Michel-Lévy chart; in comparison with the purple/indigo interference color of the accessory plate, we get 2nd–3rd order bright colors.

Countercheck: we rotate the stage 90°: now nα of the mineral coincides with γ of the accessory plate. Retardations are subtracted and interference color is lower in the Michel-Lévy chart: the result is a 1st order orange yellow or yellowish gray.

The opposite is observed in the case of a negative elongation: bright colors when the section has a NW–SE direction, pale first order colors in the direction SW–NE.

In the case of minerals with interference colors of second to third order, using a mica plate (in French "mica quart d'onde" = quarter-wave) is used; its thickness corresponds to a birefringence of a quarter of an order, so that it produces a medium gray interference color. The approach is similar to that used with an accessory plate of quartz.

2.5.5   Twinning

A twin is an intergrowth of two or several crystals of the same substance, joined by a symmetry law:

- in *contact twins* the symmetry in relation to a plane, the *twinning plane*; the contact is then a plane: for instance albite twin in plagioclases;
- in *penetration twins* the symmetry is in relation to an axis, the *twinning axis*, the two crystals are joined on an *irregular contact*: for instance Carlsbad twin in the feldspars.

Twins may by simple or repeated. Among the repeated twins it is distinguished:

- *Polysynthetic* twins: the joined individual crystals form parallel lamellae: twins in high temperature cordierite; albite and pericline twins in plagioclases
- Cyclic twins: individual crystals form a more or less circular association: twins of leucite, analcime, cordierite, etc.

Under the microscope, twins are shown by differences of interference colors of the associated crystals.

Syneusis (textures peculiar to igneous, and more commonly to volcanic, rocks where several crystals are associated in any manner in the core of the structure, the whole being surrounded by a common envelope), **Growth forms** and **deformation of minerals** have more to do with the study of the texture of the rocks than with the determination of minerals.

2.6   OBSERVATIONS IN CONVERGENT POLARIZED LIGHT

Previous observations were made with a parallel light beam. The addition of the condenser between the light source and (lower) polarizer and the thin section, transforms the light beam into a (slightly) conical beam. The observation is done with the upper polarizer (analyzer) in place.

The various narrow light beams that form this cone crosscut the studied section in different directions: they pass through *variable thicknesses* of the section and meet the index ellipsoid with a *variable incidence*. At the exit of the section, each light beam therefore presents different interference colors and intensities. The recomposition of the different light beams produces a complex interference pattern composed of:

- **isochrome lines** where the vibrations are in phase, these curves are show interferences colors in accordance with the succession Michel-Lévy chart. The isochromes can be clearly seen only if the mineral has a high birefringence (calcite for example) or if the section is (too) thick;
- **isogyres** (in French, "lignes neutres" = neutral lines) where the amplitude of vibration is null and therefore appear black. The points of emergence of the optic axes (or axis) belong to the isogyres; it is sometimes called melatope.

Whilst in the preceding observations we looked at the image given by the ocular, the obtained image is in the focal image plane of the objective. Therefore, to observe this image, we have either to remove the ocular or add a device, the Bertrand lens, used to observe the image through the ocular.

The pattern of the interference figure depends on the orientation of the section. It is only for specific sections that these figures are easily interpretable: sections perpendicular to an optic axis and sections perpendicular to the bisector of an optic axis. Do not to try to interpret a figure too far from these classic cases!

2.6.1 Obtain an interference figure

1 choose the section studied among the different sections of the mineral present in the thin section – this is the most difficult –
  - A section perpendicular to an optical axis remains always extinct in cross polarized light;
  - If you look at a section perpendicular to the bisector of the optic axes, you must have an idea about the nature of the mineral and its structure: for example, in sodic plagioclase these sections are almost perpendicular to the two cleavages.
2 Use the higher magnification so that the studied mineral occupies the entire field and perfectly focus, the objective must be perfectly centered.
3 Place the analyser, if it not already done.
4 Flip the condenser and fit it so that its lens is almost touching the section.
5 The observations are made either by removing the ocular (and possibly using an eyepiece) or by placing the Bertrand lens.

2.6.2 Uniaxial mineral: section perpendicular to the optic axis

The isogyres are two straight lines parallel to the planes of polarization of the Nicols (and so to the cross hairs of the reticle) which form a black cross, perfectly centered if the section is effectively perpendicular to the optical axis. This black cross does not move when you rotate the stage.

The isochromes are circles centered on the axis of the microscope.

If the section is not quite perpendicular to the optic axis, the black cross is not centered at the center of the field (crossing of the cross hairs) but its center is at the trace of the optic axis in the field (melotope). When you turn the plate, the trace of the optic axis describes a circle and the black cross moves parallel to itself. Strictly speaking, isochromes are no longer circles.

Such a figure may also be interpreted if the center of the cross is out of the field. Anyway, you have to be very cautious in the interpretation.

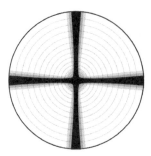

Figure 2.10 Convergent light interference figure of a uniaxial mineral cut perpendicular to the optic axis.

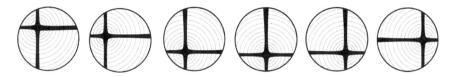

Figure 2.11 Non centered black cross.

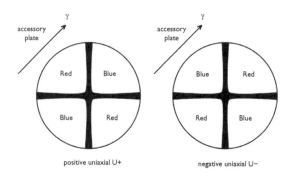

Figure 2.12 Optic sign of a uniaxial mineral.

2.6.2.1   *Determining optic sign*

Add an auxiliary plate (a quartz plate in Figure 2.12), the quadrants now show contrasting colors:

- blue in the NE and SW and orange in the NW and SE for a positive uniaxial mineral;
- the reverse: blue in the NW and SE, orange red in the NE and SW for a negative uniaxial mineral.

2.6.3   Biaxial mineral: section perpendicular to an optic axis

The isogyre is a branch of hyperbola passing through the center of the field.

If the stage is rotated, the hyperbola rotates in the opposite direction, being reduced by 4 times each turn to a straight line parallel to the cross hairs of the reticle.

The isochromes are curves, close to ovals of Descartes (actually very close to circles).

The curvature of the hyperbola is a function of 2V: very strong for small 2V, close to the curvature of the field for 2V = 45°, nearly straight for 2V close to 90°.

2.6.3.1   Determining optic sign

By adding an accessory plate (quartz in the case of Figure 2.14) sensitive, with the concavity of the hyperbola is in the NE quadrant:

- if the mineral is positive biaxial, the NE quadrant take blue colors (higher in the Michel-Lévy chart), SW quadrant takes red-orange, yellow, etc. colors;
- if the mineral is negative biaxial, the reverse occurs.

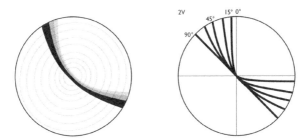

Figure 2.13 Convergent light interference figure of a biaxial mineral cut perpendicular to the optic axis.

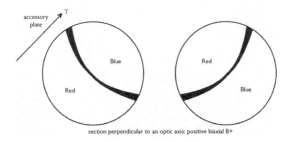

Figure 2.14 Optic sign of a biaxial mineral cut perpendicular to an optic axis.

2.6.4   Biaxial mineral: section perpendicular to the bisector of the acute angle of the optic axes

Isochromes are close to ovals of Cassini. They are centered on the trace of the optic axes in the plane of observation.

The isogyre appears as a rectangular hyperbola which admits the cross hairs of the reticle (planes of polarization) for asymptotes and passes through the trace of the optical axes. When you turn the stage, this hyperbole is deformed and degenerates into a black cross four times per turn, when the traces of the optic axes are in the planes of the cross hairs of the reticle. As the section is perpendicular to the *acute* bisector, the hyperbola remains the field.

The eccentricity of the hyperbola thus depends on 2V: it is almost cross for a small 2V, it is very open, at the limits the field of the microscope, for 2V close to 45°.

2.6.4.1   Determining optic sign

If we add an accessory plate (here quartz):

* the mineral is **positive biaxial**, γ is the bisector of the acute angle of the optical axes: NE and SW quadrants take blue colors (higher in the inter-

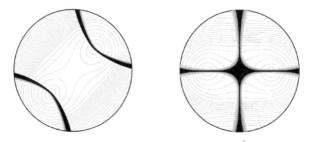

Figure 2.15 Convergent light interference figure of a biaxial mineral section cut perpendicular to bisector the optic axes.

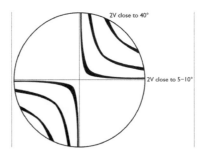

Figure 2.16 Estimation of 2V.

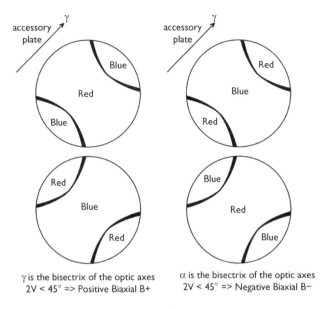

γ is the bisectrix of the optic axes
2V < 45° => Positive Biaxial B+

α is the bisectrix of the optic axes
2V < 45° => Negative Biaxial B−

Figure 2.17 Optic sign.

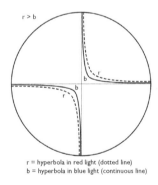

r = hyperbola in red light (dotted line)
b = hyperbola in blue light (continuous line)

Figure 2.18 Dispersion.

ference color chart) and NW and SE quadrants red-orange colors (lower in the interference color chart);

* it is the inverse if α is the bisector of the acute angle of the optical axes: mineral **is biaxial negative.**

2.6.5 Dispersion

In *some* biaxial minerals, the angle of optical axes varies with wavelength.

The figure made of convergent light on a section perpendicular to the bisector of the optical axes is a rectangular hyperbola passing through the trace of the optic axes. In dispersive minerals branches of hyperbola are more or less open depending on the wavelength:

- If 2V is larger for red light than blue/(purple) light, noted r > b, there is a red fringe in the concavity of the branches of the hyperbola, and a blue fringe on the convexity of the hyperbola:
- In the opposite case, b noted b > r or more often r < b, the color fringes is inverted.

2.7   FLUID AND MELT INCLUSIONS IN ROCK-FORMING MINERALS
By Jacques Touret

Most rocks are formed or have been influenced by a fluid phase, which may leave remnants in the form of inclusions in a number of minerals, first of all quartz. These may be of any size, from mineral defects up to huge cavities (liters in size!) contained in some pegmatite giant crystals. These inclusions, sometimes mentioned under the very improper name of «bubbles», are presently the object of extremely active research.[1] They are most (not to say only) useful in a rather restricted size range, roughly between 5 and 30 μm in diameter: smaller, they can hardly be seen under current petrographic microscope, larger, they are likely to have suffered later perturbation.

This small size requires observation under high power objectives, commonly used in routine practice for convergent figures: ×25 or ×40. It must be observed that larger inclusions (20 to 30 μm) are destroyed during the preparation of standard thin sections, only 30 μm thick. Specialized fluid inclusion studies are done on double-polished loose plates (wafers), whose thickness, depending on mineral optical quality, can be >100 μm thick. These wafers are only necessary if adapted fluid inclusion techniques (microthermometry, Raman microspectrometry) are to be used. These are extremely time-consuming and require a special expertise. But before engaging in this field of research, any petrologists should know that he can done a number of interesting observations, which will notably indicate if this type of study is simply possible: no need to call for the most advanced analytical instrumentation, if inclusions are not there. A good alternative, which I use systematically for all reconnaissance work, is to use thin sections approximately 50% thicker than usual, quartz polarizing in bright yellow.

2.7.1   Definitions

A fluid/melt inclusion is a cavity entirely contained in a mineral host, filled by a fluid phase at the time of the formation of the cavity. In fluid inclusion, this filling phase has remained fluid at room temperature, whereas in melt inclusions, occurring only in magmatic rocks, it has been transformed into a more or less devitrified glass. The overall aspect of both types is strikingly similar, to the point that it might be not easy in some cases to distinguish between both types; moreover, the distinction is only sharp to surface, effusive rocks (lavas). In deep-seated, plutonic varieties, hydrous melts may recrystallize slowly to a mineral agregate + fluid, nothing more than a special type of fluid inclusion.

2.7.2   Identification of the fluid/melt content

The identification of a fluid/melt inclusion is only possible if some gas/vapour is present in the form of a bubble (e.g. biphase (liquid/vapour) fluid inclusion. In fluid inclusion, the gas bubble is perfectly spherical, provided that it has enough space to expand, unique (only one bubble), moving freely in a liquid. In melt inclusions, several bubbles may be present, not systematically spherical (eventually elliptic in shape, due to magma flow), and, evidently, do not move. The spherical shape of the bubble is characteristic for a disorganized atomic structure (liquid or glass); inclusion shapes in minerals are always influenced by the mineral hosts, either through zigzag-shaped irregularities (mineral defects) or, when the inclusion content is thermodynamically in equilibrium with its host, as spectacular «negative crystals» (C−), the shape of the inclusion being the ideal growth form of the mineral host.

Further identification of the nature (chemical composition) of the inclusion content is highly specialized work. A first indication is given by the colour (beware of possible influence of the colour of the host), refractive index (related to the colour, see below) and, above all, the temperature at which the gas bubble disappear, either by shrinking (homogenization to liquid), expansion (homogenization to vapour), or sudden disappearance of the gas/liquid limit (critical homogenization). A precise determination requires a temperature controlled microscopic stage (microthermometry), but a first indication can be given by the temperature of the room, eventually raised on the preparation by a simple hair-dryer. If the gas bubble remains above +31°C, the fluid inside can only be aqueous, of variable salinity (would be evaluated by freezing). If the gas bubble disappear at exactly +31°C, by critical homogenization, then the fluid can only be CO_2 of critical density (0.473 g/cm^3). If the gas bubble disappears at lower temperature, in the range possibly reached in a lab (let us say >0°C), then the fluid is dominantly CO_2, low density if it homogenizes to vapour, high density if it homogenize to liquid. But it may also contain other components (CH_4 and or N_2, notably), only evidenced by chemical analysis (Raman microspectrometry). These rough estimates are somewhat enhanced by the colour of the different phases: The refractive index (N) of gases is uniformly 1, resulting in a dark, almost black colour under the microscope. Liquid CO_2 (N = 1.11) is distinctly darker than liquid H_2O (N = 1.33).

Inclusions, either fluid of melts, can also contain solid minerals often referred to as «daughter» phases. Their identification is a major aspect of any inclusion studies, not easy in reason of their small size. Again a highly specialized work, especially as spot analyses are commonly not possible, because of the mobility of the solids within the enclosing liquid. Optical characteristics are the first parameters to evaluate (see Table), as well as few general remarks:

i   Except for H_2O-soluble species (halides) (very common in aqueous fluids), many «daughter» phases a can also be s commonly found in the groundmass, notably as mineral inclusions,

ii Many daughter phases are well crystallized, with well-developed crystalline faces. These are sometimes sufficient to identify the mineral species,

iii When several daughter phases are present (up to about 30 in some topaze from Volhyn, Ukrainia!), each crystal corresponds to a different mineral species. If they fill most of the cavity, a good trick may be to freeze a section or a mineral grain in liquid N_2, break it and put the broken face as fast as possible in EDS-equiped Scanning Electrom Microscope. Not easy, requires an extremely skilled technician, but not impossible, as I have experienced in Amsterdam.

Most difficult inclusions to identify are high-density single phase gazeous, like those found in high-pressure/high-temperature metamorphic rocks (granulites/eclogites). Their content homogenize at very low temperature (less than $-143°C$ for N_2 inclusions in eclogites) and, until this temperature, they cannot be distingusih from empty cavities (holes), present in virtually any rock section, filled with air at atmospheric pressure. Experience can be of some help, as well as very simple crushing tests under the microscope (crushing a small grain in a droplet of oil or glycerine between two glass plates) releases large gas bubbles.

2.7.3   Primary versus secondary inclusions

A most important, much debated aspect of fluid inclusion studies is the timing of inclusion formation in respect to its mineral host. Primary inclusions are made at the time of the mineral growth, whearas secondary inclusions may be formed much later (sometimes million years later!), in healed microfractures. Detailed criteria are found in the inclusion literature. It will be sufficient here to say that primary inclusions are either isolated, corresponding to perturbations during the mineral growth or, e.g. in «phantom» idiomorphic crystals, aligned along growth structures. Secundary inclusions, on the other hand, are distinctly aligned and regularly spaced along trails, not immediately related to the host mineral structure (except in the case of easy cleavage). Further identification is outside of the scope of the present work but few remarks are appropriate here:

i   Primary inclusions are not the «good», secundary the «bad» inclusions. A number of perturbations may occur during or after the formation of a primary inclusions, e.g. cavity caused by sticking of a gas bubble on a grawing face or leakage, making the inclusion content quite different from the interstitial fluid phase at the time of the mineral growth.

ii   Conversely, secondary inclusions might be quite representative, if P – T conditions have not changed substantially since the mineral growth. But even if this has happened, knowing their existence and characteristics can be quite significant: a rock is never traversed by percolating fluids without important petrographical or geochemical consequences.

In conclusion, each inclusion category must be identified, then studied as a single population, applying the only criteria which, in each case, guarantee the representativity of the inclusion fluid: comparison of P – T fluid data with independant mineral estimates (see fluid inclusion literature). As long as an unambiguous conclusion has not been reached, it is important to stick to a proper langage: primary versus secondary for inclusions distinctly related to the host mineral gowth, neutral terms (early verus late) if not.

2.7.4 Potential interest of fluid/melt inclusion studies

With all their limitations in mind (difficult of observation measurement, mutiple possibilities of perturbations, etc..), it must be recognized that fluid inclusions are the only possibility to record phases having left the rock system. Furthermore, they are also the only objects likely to inform, not only of the chemical composition, but also on the molar volume (density) of a fluid system. A variable of state most commonly ignored by petrologists, because of the small variation for solid minerals in the P-T range of geological interest, but of tremendous importance for fluids: a mole of CO_2 has a volume of less than 20 cm^3 during granulite metamorphism at depth, against more than 22000 cm^3 when it reaches the Earth's surface. Once an inclusion is sealed, its volume as well as the mass of the enclosed fluids are fixed, hence its molar volume (PVT equation of state), which will no vary, provided that no further perturbation will occur. Again, a complete investigation is out of the scope of the present work. But a preliminary examination, which should be done by any petrologist during routine microscopic observation, should provide some of the following results:

i   Identify «good» samples, namely those containing, at least locally, a great number of workable inclusions. … examples

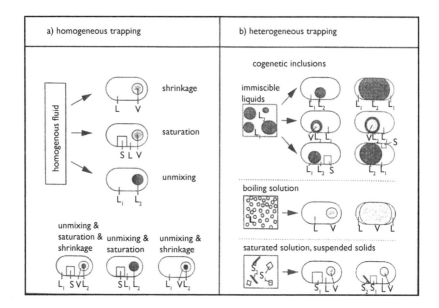

ii   In these samples, identify the different inclusion-containing minerals. If the rock contains quartz, this mineral will in principle contain most inclusions. This gives a first indication (if quartz does not contain workable inclusions, better do something else. But quartz traps easily many inclusion generations, which might be quite difficult to study properly. If inclusions are found in other minerals (apatite, garnet, pyroxenes, feldspars), then their study might be much more simple and rewarding.

iii Experiences shows that, in most rocks, the apparent complexity of millions of inclusions resolves in a small number (typically less than 5, most commonly 1 or 2) of fluid types, a fluid type being defined by a relatively constant chemical composition and a limited variation in density (for details, see the fluid inclusion literature).

iv   Once a fluid type has been identified, most important is to evaluate its homogeneity or heterogeneity (described in the inclusion literature as homogeneous *versus* heterogeneous trapping, see Table below (from Van den Kerkhof & Hein, 2010). For aqueous fluids, the only ones which can be studied for this purpose without specialized equipment, homogeneity is immediately indicated by a constant volume ratio of the different phases: degree of fill (Liquid/total) (for pure water giving immediately the density), solid/liquid, etc... If the timing of the fluid trapping is known, then the interpretation of homogeneous fluid system is immediate, in first approximation (homogenization temperature for H_2O evaluated from the degree of fill).

v Heterogeneous fluids are much more complicated to analyze, better make any conclusion without additional data. They may result from perturbation (leakage), successive fluid pulses or, again a result that only inclusion can provide, fluid immiscibility. Be careful, not to jump immediately to this tempting conclusions: I have seen a number of «boiling» examples (liquid/vapour immiscibility) simply due to liquid H_2O leakage.

vi Highly complicated (but also very important) to be evidenced for fluids, immiscibility is far more simple – but equally important – for melt inclusions. These require for their proper investigation special equipment (high-temperature heating microscopic stages), only to be found in a small number of highly specialized lab. But the two basic type of melt immiscibility (melt/melt or melt/fluid) are easily identified under the microscope, always from the same criteria (droplet of A within B, see illustrations). The fact that it is possible to bring the glass inclusion at the thin section surface for microporbe analysis is an additional advantage. But, again do not believe that a complete interpretation is easy work. If you observe these phenomena in your samples, they are indeed interesting rocks. But if you see them for the first time better show them to a specialist!

Chapter 3

Systematic mineralogy

3.1 MAJOR TECTOSILICATES: QUARTZ – FELDSPARS – FELDSPATHOIDS

In tectosilicates (= framework silicates), tetrahedra $(SiO_4)^{4-}$ form a three-dimensional lattice in which each oxygen placed at an apex of a tetrahedron is shared between two neighboring tetrahedra. The silicon Si^{4+} ions in the center of each tetrahedron can be replaced by aluminum Al^{3+} ions. All the aluminum in the tectosilicates is 4-fold coordinated. Two neighboring tetraedra sharing an apex cannot contain both Al^{3+} ions: the replacement of Si by Al can not exceed 50%. This substitution of Si by Al can be completely disordered or partially ordered. Disordered forms whose entropy is higher, are high temperature polymorphs.

The deficit of the charge between Si^{4+} and Al^{3+} ions is compensated by large cations K^+, Na^+, Ba^+ and Ca^{++} which are 6-fold or more coordinated. Small and/or heavily charged cations induce a local imbalance and do not fit into these structures: tectosilicates do not contain ions such as Fe, Mg, Mn, Ti. Due to the larger dimensions of their cations, lead or Rare Earths can enter these structures.

The structure of tectosilicates is relatively open and molecules of large dimension, Cl_2, CO_3, SO_4, S_2 can enter the lattice (scapolite, feldspathoids of the sodalite group). Zeolites contain open cavities where water molecules can be introduced; this water is very weakly bound to the lattice.

As Si is replaced of by Na Al or K Al, or 2 Si by Ca Al_2, all tectosilicates are strictly saturated in alumina.

Tectosilicates include:

1 silica group minerals;
2 feldspars;
3 feldspathoids that are under-saturated in silica in comparison to feldspars;
4 analcime that is not a feldspathoid but is close to them by its chemical composition and its occurrences;

5   scapolite group, that, due to their occurrences, will be treated in this
    book with the calcic minerals;
6   zeolites group (not treated here).

3.1.1   Silica group

There are six natural mineral with SiO_2 formula, three of them have high
and low temperature polymorphs:

Quartz: trigonal α-quartz is stable up to 573°C; hexagonal β-quartz is
stable from 573°C to 870°C.

Chalcedony is a fibrous habit of low temperature quartz. The fibers are
elongated in the c direction. *Quartzine* has a positive elongation and *chalcedony* a negative one. *Lutecite* is a fibrous habit of low temperature silica
in which the fibers are elongated obliquely on the c axis.

Tridymite: orthorhombic α-tridymite is an unstable polymorph that can
exist up to 117°C; hexagonal β-tridymite is unstable from 117°C–870°C
and stable from 870°C to 1470°C.

Cristobalite: quadratic α-cristobalite, there may exist up to about
200–275°C. It is unstable at these temperatures. Cubic β-cristobalite exists
above 200–275°C. It is stable only at temperatures above 1470°C, its melting point is 1713°C.

Coesite is a monoclinic polymorph, stable at pressures exceeding 28 to
30 kb.

Stishovite is a quadratic polymorph of ultra-high pressure.

Moganite (Florke et al., 1976, 1984, approved in 1999) is a monoclinic pseudohexagonal polymorph of silica of low-temperature. Moganite
is metastable and recrystallizes into quartz.

Quartz is, after the feldspars, the most common mineral in the Earth's
crust forming 10–15% of it. It occurs in all families of rocks and its stability
field covers almost the entire field of the geological phenomena

In *sedimentary rocks*:

Because of its mechanical and chemical resistance, quartz is a abundant clastic mineral. Sedimentary sorting, especially concentrates quartz, in
sands and sandstone of various types. In these rocks, quartz can be corroded or otherwise preexisting grains can be nourished by secondary quartz.
The quartz is also a common cement in sedimentary rocks. The newly
formed quartz is common, either as euhedral crystals (authigenic quartz)
or as fibrous aggregates (chalcedony). Some siliceous rocks such as chert,
buhrstone (= "meulière"), etc. consist of aggregates of chalcedony that leave
besides abundant micropores.

Opal is a form of silica that is either micro-crypto-crystalline, cristobal-
lite (possibly with tridymite) (C opal) or aggregates of colloidal silica (opal
A), that contains up to 20 wt% of water and the bulk composition SiO_2, n
H_2O. Opal is the form of silica that occurs in the skeleton of living creatures

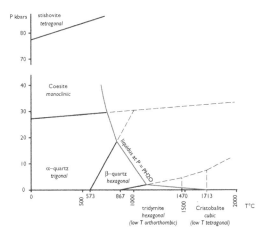

Figure 3.1 Polymorphs of silica (from Tuttle and Bowen, 1958, modified).

(diatoms, radiolarians, siliceous sponges) and in the sediments derived from them. Cristobalite is unstable at room temperature, opal tends to transform into fibrous quartz (chalcedony and lutecite); this transformation is very slow.

The solubility of cristobalite in the surface waters is 120 ppm; the one of quartz is 7 ppm, it is therefore the main source of silica in solution. The lutecite seems crystallized directly from water at high pH, particularly in an evaporitic environment.

In *hydrothermal rocks*, opal precipitates from waters at temperatures lower than 100–150°C. At higher temperatures it is chalcedony or quartz (α-quartz) that precipitates. In the veins and particularly ore veins, quartz often occurs in elongated prismatic crystals that tend to grow perpendicular to the walls (comb structure).

In most *metamorphic rocks* the stable form of silica is quartz: α-quartz or β-quartz. The latter is transformed into α-quartz during the fallout of temperature. In the ultra-high pressure facies (white schist), the stable form is coesite; it largely tends towards retrograde quartz at the pressure drop.

Coesite and stishovite are found in meteorite impacts, and, for coesite, atomic bombs impacts. Coesite also occurs in the enclaves uplifted by magmas of very ddep origin, especially in kimberlites.

The conditions of formation of *plutonic rocks* are the same as that of metamorphic rocks (possibly with higher temperatures: 650–1100°). Quartz crystallizes when the silica content of the rock is high enough, this limit depends on the composition of the rock: it is about 50 wt% in mesocratic rocks, 60 wt% in the leucocratic rocks. The stable form of quartz is β-quartz; inversion into α-quartz produces characteristic fractures in it. Quartz in igneous rocks is generally anhedral. In some rocks, especially in hypabyssal rocks, it shows a habit of dipyramid with or without a very short prism. *Graphic textures* result from the intergrowth of quartz and potassium feldspar. *Granophyric textures* (also called micropegmatite) are the result of dendritic growth of quartz around early phenocrysts in the microcrystalline groundmass of some porphyritic rocks. *Myrmekites* are bud- or wart-like structures, with vermicular intergrowths of quartz in acidic plagioclase, that develop in the potassium feldspar at the contact between a plagioclase and a potassium feldspar. Myrmekites are the result of replacement of K in the potassium feldspar by Na and Ca (from the plagioclase) with release of silica.

In *volcanic rocks* quartz is the silica polymorph which crystallizes in the early stages of evolution of magma at depth. It often occurs in euhedral bipyramids. These early crystals often show embayments that make their "normal" euhedral boundaries incomplete; such textures are interpreted either as dendritic-like crystallization, or corrosion of early crystals by the magma, as temperatures at the end of crystallization of the magma at the surface are higher than in depth due to oxidation reactions and the lower pressure. Tridymite may then crystallize into microliths in the groundmass of the rock during the late stages. Tridymite and cristoballite crystallize in vugs in vesicular lava or in veins in post-magmatic conditions under the effect of fumaroles.

3.1.2   Feldspars

3.1.2.1   *Chemical composition*

Feldspars may be described as solid solutions between three end members:
$KAlSi_3O_8$ potassium feldspar (K-feldspar)
$NaAlSi_3O_8$ albite
$CaAl_2Si_2O_8$ anorthite
There are two series of solid solutions:

1   A solid solution (alkali feldspar) between potassium feldspar and albite by substituting K ⇔ Na substitution.
2   A solid solution (plagioclases) between albite and anorthite, by substituting Na Si ⇔ Ca Al^{IV}. The composition of plagioclase is expressed by the percentage of its anorthite contain (An%).

Celsian is the barium end member $BaAl_2Si_2O_8$. Hyalophane form a continuous series between this end member and potassium feldspar.

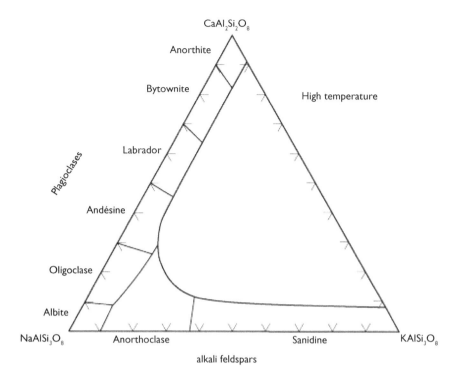

High temperature

alkali feldspars

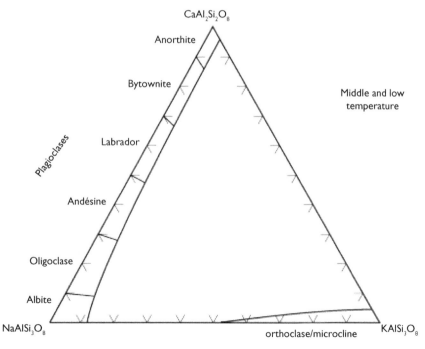

Middle and low temperature

Figure 3.2 Feldspar nomenclature.

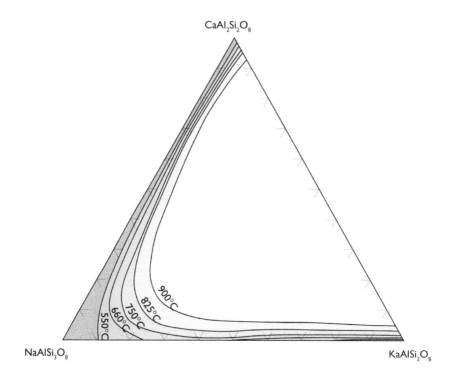

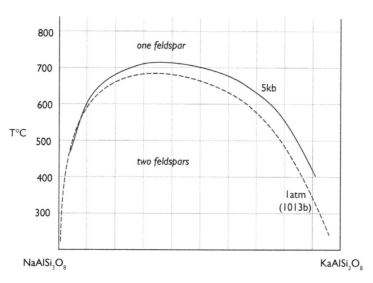

Figure 3.3 Solvus of the feldspars (modified after Fuhrman and Lindsley, 1988, Roux 2000, unpublished data, and Parson, 1978).

At high temperature (above 700°C), the series of alkali feldspars and of plagioclases are continuous and each of them may contain very limited quantities of the third end member.

At medium and low temperatures, plagioclase form a complete solid solution series. But there is a miscibility gap between K-feldspar and albite, and this gap is more important as the temperature is lower. So that when alkali feldspar has crystallized at a sufficiently high temperature and it cools slowly, exsolution of one mineral appears in the other: *perthite* is a dominant potassium feldspar with exsolutions of albite; *antiperthite* is a plagioclase feldspar with potassium feldspar exsolution, *mesoperthite* is a feldspar where the proportions of the two end members are equivalent.

At low temperatures (below 550–600°C), there is also a gap of solubility in the plagioclase between An5 and An15 (An1 and An25 below 200°C), these demixed plagioclases, forming very fine lamellar associations, are called *peristerite*. There are also two other unmixing solvi: one between An39–48 and An53–63 (Bøggild exsolutions), the other one between about An55 and An95 (Hüttenlocher exsolutions). Peristerite, and the other exsolutions, cannot be distinguished under the microscope.

In each series, there are high, medium and low temperature forms.

The potassium feldspar has three polymorphs:

1   *sanidine*: a monoclinic, completely disordered polymorph of high temperature; it occurs only in volcanic rocks.
2   *orthoclase*: a monoclinic polymorph of intermediate temperature.
3   *microcline*: a triclinic polymorph of low temperature.

Adularia is a variety of orthoclase, crystallized at low temperature in a hydrothermal environment.

Albite and plagioclase are triclinic. Albite has a form of high temperature and one of low temperature. Plagioclases have many polymorphs; they depend not only on temperature of formation, but also their anorthite content and their thermal history. Anorthite is completely ordered. The various polymorphs can not be distinguished under the microscope and their distinction requires the use of X-ray diffraction.

3.1.2.2   Stability of feldspars

Due to their open structure feldspars are not stable at high pressures and are stable at low and medium pressure in a wide temperature range.

K-feldspar is stable at ordinary temperatures. It has an incongruent melting producing leucite + a liquid at 990°C at ordinary pressure. Its melting point is significantly lowered by high water pressure. For water pressures greater than 2.5 kb, the melting becomes congruent. At (water) pressures over 10 kb, K-feldspar is replaced by muscovite (actually by phengite).

At ordinary temperatures, in the presence of water and under low pressures, albite is metastable; the stable form is analcime (analcime + quartz = > albite + water). Towards high pressures, albite is replaced by jadeite (albite = > jadeite + quartz). An albite melts in dry conditions at temperatures of about 1100°C. The presence of water significantly lowers its melting point: 748°C under 5 kb of water pressure.

Anorthite is not stable at low temperature (below 500°C). It is replaced at low pressure, in the presence of water, by zeolites: in particular heulandite. ($CaAl_2Si_7O_{18} \cdot 6H_2O$), laumontite ($CaAl_2Si_4O_{12} \cdot 4H_2O$) and wairakite ($CaAl_2Si_4O_{12} \cdot 2H_2O$) At higher pressure (5–6 kb), it is replaced by lawsonite ($CaAl_2Si_2O_7(OH)_2 \cdot 2H_2O$), and at even higher pressure (and temperature) by the association kyanite (Al_2SiO_5) + zoisite ($Ca_2Al\ Al_2Si_3O_{12}(OH)$) or kyanite + grossular ($Ca_3Al_2Si_3O_{12}$). It melts at 1550°C in dry conditions; its melting point is also greatly reduced by water pressure.

The different end members of the feldspars are rarely alone in a rock and their melting is actually that of the liquidus in the $KAlSi_3O_8$ – $NaAlSi_3O_8$ – $CaAl_2Si_2O_8$ system (or in derived binary systems), these systems are strongly dependent on the pressure water.

3.1.2.3 Occurrences of feldspars

Feldspars are the most abundant minerals in the Earth's crust, forming about 60% of it.

Sedimentary rocks

Although it is a fragile mineral and do not support long transport, feldspar is a constituent of certain clastic rocks (arkose, greywacke). K-feldspar is stable in sedimentary environment. It forms either clastic fragments or authigenic crystals (often crystallizing from clastic fragments) Albite is metastable but appears frequently as authigenic crystals. It is also described in some pedogenic alterations. Plagioclases are all the more rapidly altered as they are more calcic.

Hydrothermal rocks

Albite and adularia are common in hydrothermal veins. Calcic plagioclase never occurs in such rocks.

Albite is also formed by metasomatism of various rocks and minerals: the result of such process is a rock formed by more than 95% of albite, albitite. The albite formed at the expense of K-feldspar present, the particular habit of chessboard albite.

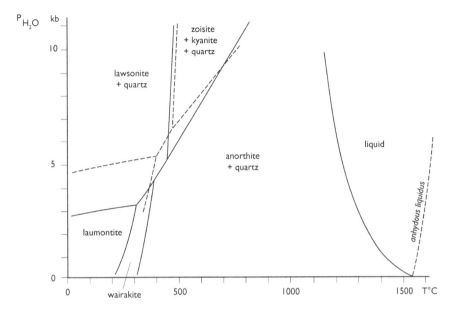

Figure 3.4 Stability of anorthite equilibria anorthite – laumontite – lawsonite – zoisite + kyanite: solid line from Newton and Kennedy (1963), dotted line from Crawford and Fyfe (1965) and Perkins et al. (1980).

Calcic plagioclase of some lavas is converted into albite by oceanic hydrothermal metamorphism (such rocks were previously called "spilites" and "keratophyres"). This phenomenon also occurs in some continental lavas.

Metamorphic rocks

Albite, calcic plagioclases and the minerals that replace them at low temperature and high pressure are index minerals of the metamorphic facies.

Albite and plagioclase are absent from the low temperature and high pressure/(water pressure) metamorphic facies. Albite appears in the greenschist facies, epidote-amphibolite facies and part of blueschist facies. Calcic plagioclase is absent from these facies, being replaced by (iron-poor) epidote. Calcic plagioclases are characteristic of amphibolite facies and granulite facies.

In metamorphic rocks, plagioclase often shows a reverse zoning with a border more calcic than the core: in prograde metamorphism, the calcic plagioclase gradually becomes progressively stable as the temperature increases. Plagioclase of metamorphic rocks that are not of igneous origin, commonly remains without twinning.

The transition from amphibolite facies to granulite facies is marked by the destruction of calcic amphibole into calcic plagioclase + orthopyroxene. There may be an intermediate stage the formation of cummingtonite (+ calcic plagioclase) and then transformation of all the amphiboles into orthopyroxenes. These various reactions are continuous reactions.

Plagioclase also appears at retrograde metamorphism from eclogite facies into amphibolite facies: omphacite (sodium-rich pyroxene) + (pyrope-rich) garnet are transformed into kelyphite, an association of plagioclase + hornblende. There may be an intermediate stage with plagioclase + diopside.

Calcic plagioclase is an important component of impure metamorphic carbonate rocks: impure marbles and calc-silicate-gneisses.

In a metamorphism of relatively low grade, microcline is the stable form of potassium feldspar: it is commonly found in the orthogneisses (plutonic or volcanic origin) and some metamorphosed clastic rocks (meta-arkose, for example). There is an isograd microcline => orthoclase which corresponds to a temperature of about 600 to 650°C.

Potassium feldspar is also formed by destruction of mica:

• the isograd muscovite + quartz = > sillimanite + potassium feldspar + water corresponds to the disappearance of primary muscovite; this is an important boundary within the amphibolite facies;
• the reaction biotite + quartz => orthopyroxene (or garnet depending on the alumina content of the initial biotite) + orthoclase + water appears, approximately, at the beginning of the granulite facies. As ferromagnesian minerals are involved, it is a continuous reaction that occurs over in a certain range of temperature (*i.e.* some extent of field).

Igneous rocks

The classification of igneous rocks is based on the relative proportions of alkali feldspar/plagioclase/quartz/feldspathoids. This shows the importance and abundance of feldspars in these rocks.

Since the work of Bowen (1915/1928) and Tuttle and Bowen (1958), crystallization of feldspars has been the subject of numerous studies of experimental petrology. They are widely exposed in the courses of petrology. They also clarify the critical role played by the water pressure in the magmatic crystallization of feldspars.

Binary systems: albite (Ab) – anorthite (An) system

Consider a magma evolving at decreasing temperature under low water pressure, it reaches the liquidus in L1 and crystallizes plagioclase F1. The liquid then evolves from L1 to L2 and plagioclase shows normal zoning F1–F2.

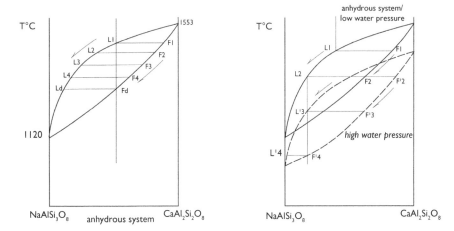

Figure 3.5 Albite – anorthite system (after Tuttle and Bowen, 1958).

In practice, we commonly observe oscillatory zonations and even (partially) reverse zonations. This indicates that the system has not crystallized in a closed system. Variations in water pressure, in particular, induce such reverse recurrent zoning (Figure 3.5). Suppose that at L2, occurs an increase in water fugacity (high f_{H_2O}). Liquidus and solidus are lowered so that the plagioclase in equilibrium with the liquid L2 will be F'2 plagioclase that is much more calcic than the plagioclase F2. Strictly speaking, if the change in water fugacity is sudden, the liquid L2 is no longer at the liquidus and resorption of the already crystallized plagioclase may occur. The temperature must again decrease so that this liquid reaches in L'3 the liquidus at high water pressure: it then F'3 crystallizes feldspar. In fact, during the fractional crystallization of magma, water fugacity increases gradually producing a reverse zoning of the plagioclase which becomes more and more calcic toward its rim. Then, the evolution of the liquid occurs along the liquidus, thus producing a normal zonation F'3–F'4. On the other hand, abrupt departure of fluid, and thus abrupt decrease of water pressure, can occur during the crystallization of magma.

*Binary system of alkali feldspars: albite (Ab) –
potassium feldspar (FK) system*

Under low water pressure, the field of the liquid (:melt) in the phase diagram of the system albite – potassic feldspar is entirely located above the solvus. Such evolution is called *hypersolvus* crystallization. In this binary system, the albite that crystallizes, is enriched in K-feldspar constituent; If potassium feldspar crystalizes, it is enriched in Na-constituent. Fractional crystallization produces

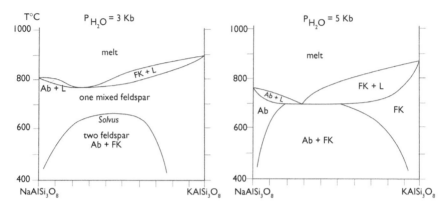

Figure 3.6 Albite – potassium feldspar system (after Tuttle and Bowen, 1958).

a mixed sodi-potassic feldspar. This is what usually occurs in evolved volcanic rocks like phonolites, trachytes and many rhyolites.

The presence of water significantly lowers the crystallization temperature so that the liquidus and solidus intersect the solvus. This evolution is called *subsolvus* crystallization. Then there is crystallization of two feldspars, a sodic feldspar (plagioclase) and a potassic one. Most granitoids magmas show subsolvus evolution and crystallize two feldspars at the magmatic stage. Hypersolvus granitoids are very rare.

Ternary system anorthite – albite – K-feldspar

The crystallization of feldspar in magmas more calcic than granitic magmas cannot be properly represented by the system albite – K-feldspar. So the full ternary diagram albite – potassic feldspar – anorthite must be used.

This highly complex diagram must also represent the liquidus, solidus and solvus. We will break it down here into:

A *first diagram* showing the solidus and solvus under high water pressure; the intersection of the solvus and solidus divides the field of feldspars into two disjoint domains: plagioclase and alkali feldspar.

Under lower water pressure, the solidus is at a higher temperature and intersects only partially the solvus: there is a continuous domain for a mixed single feldspar near the anorthite–albite and albite–K-feldspar tie lines: this is the case of hypersolvus crystallization.

A *second diagram* shows the liquidus materialized by the isotherms: there is a cotectic valley between the fields of plagioclase + liquid and liquid + alkali feldspar. Temperature decreases along this cotectic line from the point E, the An–FK tie line, toward the F eutectic point (close to the Ab–FK tie line).

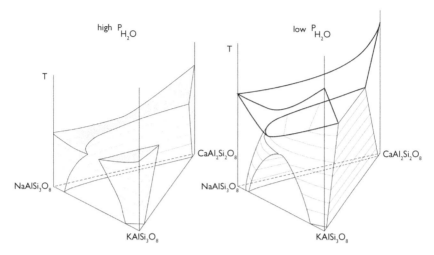

Figure 3.7 Solidus and solvus of the feldspars.

Left, diagram under a high water pressure: solvus and solidus intersect so that the stability field of feldspars forms two disjoint volumes (materialized by the isotherms). They are limited by the solvus and toward high temperatures, by the solidus (tight hatch).

Right, diagram under low water pressure: the boundaries of the solidus are represented by a bold line. The dome materialized by isotherms represents the two-feldspar field. The volume between the two surfaces is the domain of mixed feldspar.

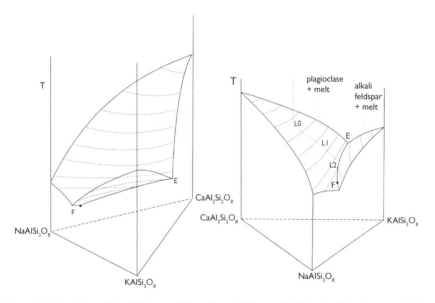

Figure 3.8 Liquidus of the feldspars (after Best 1982, modified) It is difficult to read this diagram if it is oriented in the same way as the previous diagrams (left figure), so it has been differently reoriented on the right figure.

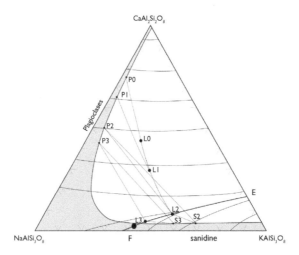

Figure 3.9 Crystallisation of the feldspars (after Best and Christiansen 2001, modified).

In projection on the plane albite – anorthite – k-feldspar (Figure 3.9), the liquidus is represented by isotherms and its intersection with the solvus; the shaded area represents the field of liquid + one single feldspar (under low pressure water in the case shown).

Consider a magma that reaches the liquidus in L0 (Figure 3.8), this magma crystallizes plagioclase P0, more calcic than the initial liquid (Figure 3.9). Then during the cooling, the magma evolves along a line L0–L1–L2, plagioclase by P0–P1–P2, and becomes more and more sodic.

The K-feldspar content of the liquid and of the plagioclase increases until reaching the EF cotectic line in L2. There is then simultaneous crystallization of P2 plagioclase and S2 K-feldspar. Then the liquid follows cotectic line from L2–L3 crystallizing P2–P3 plagioclase that becomes increasingly sodic (and potassic) and S2–S3 K-feldspar that becomes increasingly sodic. In this simple case the magmatic plagioclase shows a normal zoning from a more calcic center to a more sodic rim; the K-feldspar shows a zonation with a more potassic center to a mixed more sodic rim.

The cotectic line obviously stops at the boundary of the domain of single feldspar. But crystallization may extend into this area until the minimum temperature ("eutectic") F point. In evolved volcanic rocks (phonolitic, trachytes and rhyolites some) a single (mixed) feldspar then crystallizes: the liquid has passed the final point of the cotectic line. In contrast, in plutonic rocks that crystallized at lower temperature, the point F is located in two feldspars field: two feldspars crystallize until the end of the evolution; this is a subsolvus evolution.

In plutonic rocks, plagioclase crystallizes in a low temperature form and potassium feldspar is orthoclase or microcline. In the rocks that have crystallized at relatively low temperatures (granites, granodiorites. etc.), solid solutions between K-feldspar and plagioclase remain limited and slow cooling leads to exsolutions, observable under a microscope, the *perthites.* In other rocks, especially the alkaline rocks (syenite, nepheline syenite, etc.) solid solutions between K-feldspar and plagioclase are much larger and exsolution leads to mesoperthites. Perthites show different habits: from very fine and regular *film perthites* since the, to thicker, irregular *rods perthites.* This habit depends on the initial proportion of K-feldspar and albite in the initial mineral and of the cooling history. *Antiperthites* occur in plagioclase of plutonic rocks that crystallized at high temperature under conditions of granulite facies (charnockite, norites, etc.).

Graphic textures are mainly observed in pegmatites. They result from a syncrystallization quartz and feldspar in the late magmatic evolution.

Rapakiwi textures appear in the development of albite around K-feldspar phenocrysts. The albite may be monocrystalline or polycrystalline, and it sometimes forms radial aggregates. There may be alternating layers of albite and K-feldspar. Interpretations call upon surface phenomena and mainly to changes in water pressure that shift slightly the liquidus.

Myrmekites are bud-or wart-like textures of acidic plagioclase containing quartz vermiculi developed at the contact between K-feldspar and plagioclase from the plagioclase into the K-feldspar. Myrmekite are interpreted by a subsolidus reaction: Na and Ca diffuse from plagioclase towards K-feldspar and replace potassium in the K-feldspar; the latter is then transformed into plagioclase with release of silica (quartz).

In volcanic rocks plagioclase are high temperature forms and alkali feldspar are sanidine or anorthoclase. In the latter the exsolutions remain invisible under the microscope and these cryptoperthites can only be detectable by X-ray. Feldspars are strongly zoned and show more numerous and varied twinnings than in the plutonic rocks due to faster cooling. In evolved rocks (phonolitic, rhyolites, trachytes some) there is a single (mixed): feldspar: the liquid has passed the final point of the cotectic line.

3.1.2.4 Alteration of the feldspars

Red clouding is a common and distinctive alteration of orthoclase: the feldspar is commonly stained into brown or reddish brown by a pervasive development of a very fine grained mineral which cannot be determined under the microscope: it may be kaolinite or sometimes hematite.

Potassium feldspar, orthoclase and microcline, are altered into muscovite by a leaching of the alkalis. The microscopic appearance of this muscovite is very different from the alteration of the plagioclase into sericite:
muscovite shows a habit of large crystals with very irregular outlines, called
mottled muscovite, or sometimes «spongy» or «skeletal» muscovite, it is
developed in the fractures or at the rim of the feldspar, the rest of which
remains unaltered.

Sericitization is an alteration of plagioclase in a fine muscovite, sericite.
In comparison with muscovite, sericite often has a lower content in potassium and a higher one in silica and thus is close to illite. Sericite forms very
small dispersed crystals, oriented or not, affecting the whole feldspar. Sericitization often develops being guided by the structure of plagioclase: cleavages, zoning, etc.; the more calcic parts of plagioclases, especially the core of
crystals, are more extensively sericitized.

Saussuritization in a replacement of calcic plagioclase by albite + epidote
$Ca_2(Al,Fe^{3+})Al_2(Si_2O_7)(SiO_2)(OH)$ (and sometimes calcite, sericite, zeolites).
Saussuritization reflects a leaching of calcium, addition of water (and ferric
iron), oxidizing conditions and a fall in temperature. It is a retromorphosis
into greenschist facies. This hydrothermal alteration is often associated with
ore deposits.

The albitization is a metasomatic transformation that develops both on
potassium feldspar (chessboard texture) and plagioclase.

Feldspars can also be altered into chlorite, pumpellyite, prehnite, some
zeolites, etc.

3.1.3   Feldspathoids

Feldspathoids are tectosilicates close to the feldspars but differ by a lesser
amount of silica. They are thus incompatible with the quartz, orthopyroxene and pigeonite. Some feldspathoids contain salts: chlorides, sulfates, carbonates, sulfides. Analcime is not a feldspathoid but contains zeolitic water,
it is close to feldspathoids by its chemical composition and its occurrences.

3.1.3.1   Chemical composition

Feldspathoids are classified according to the Si/Al ratio:

- Si/Al = 2: leucite group, minerals equivalent to a feldspar *minus 1*
  SiO_2:
  - **Leucite** $KAlSi_2O_6$ (tetragonal pseudocubic)
  - **Analcime** $NaAlSi_2O_6 \cdot H_2O$ (zeolitic water) (cubic)
- Si/Al = 1: minerals equivalent to a feldspar *minus 2* SiO_2:
  - without additional salts (hexagonal minerals):

- **Nepheline** $Na_3(K, Na)(AlSiO_4)_4$ $[NaAlSiO_4]$ There are three polymorphs: a low temperature hexagonal polymorph (ordered, stable up to about 900°C), a high-temperature hexagonal polymorph (disordered, up to 1254–1280°C) and a cubic polymorph, carnegieite (unknown in nature).
- **Kalsilite** $KAlSiO_4$ With numerous poplymorphs, and among them kaliophilite.
  - with additional salts:
    - sodalite group (cubic minerals)
      Sodalite $Na_8(AlSiO_4)_6Cl_2$
      Noseane $Na_8(AlSiO_4)_6SO_4$
      Haüyne $(Na, Ca)_{4-8}(AlSiO_4)_6(SO_4, S)_{1-2}$ series
    - **Cancrinite** ($CaCO_3$ end member) – **vishnevite** (Na_2SO_4 end member) series (hexagonal minerals) $(Na, Ca, K)_{6-8} (AlSiO_4)_6 (CO_3, SO_4, Cl)_{1-2} (H_2O)_{1-5}$

Most natural nephelines have compositions close to $Na_3 K(AlSiO_4)_4$. Substitution K ⇔ Na is generally limited because these elements occupy different sites: Na is 8-fold coordinated, K is 12-fold coordinated. Experimentally, there are solid solutions (in both nepheline and carnegieite) up to a pure sodic end member $NaAlSiO_4$. The nepheline from some ultra-potassic lavas of high temperature show solid solutions between nepheline and kalsilite up to $Na_{2,3} K_{1,7}$ compositions.

Substitution Na Al ⇔ Si □ (the symbol □ means vacant site) occurs in low proportion in the high-temperature nepheline, indicating a limited solid solution of alkali feldspar in nepheline.

Calcium enters in very low proportion in the nepheline by Na Si ⇔ Ca Al substitution.

Substitution K => Na occurs only in very rarely in kalsilite; practically never in leucite.

3.1.3.2 Occurrences

Feldspathoids are minerals of very deficient in silica rocks, mostly igneous rocks.

Nepheline is stable under relatively high water pressures: it appears as well in volcanic rocks (high temperature polymorph) as in plutonic rocks (low temperature polymorph). It coexists with albite and acidic plagioclase and thus appears in basic as well as evolved rocks.

Sodalite is a mineral of plutonic and volcanic rocks, but rather appears in differentiated rocks: nepheline syenites and their volcanic equivalents.

Nosean and haüyne appear only in volcanic and hypabyssal rocks of the phonolites family.

Leucite occurs exclusively in very potassic, mostly basic volcanic (and hypabyssal) rocks. In the presence of albite it is transformed into

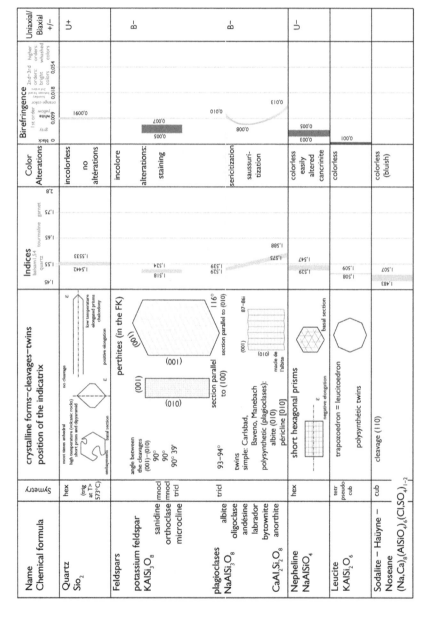

Figure 3.10 Summary of the characters of quartz, feldspars and feldspathoids.

sanidine + nepheline; that explains its absence in evolved rocks. Sometimes early crystallized leucite is destabilized in a very fine aggregate of sanidine + nepheline which retains its crystalline forms (pseudo-leucite). Leucite is unstable under high pressure, which explains its absence in the plutonic rocks.

Kalsilite is a very rare mineral of ultrabasic potassic volcanic rocks.

Cancrinite is essentially a secondary mineral formed on nepheline.

Nepheline, sodalite (and cancrinite) appear by metasomatism (fenitization) of very varied rocks (gneiss, amphibolite, limestone) around alkaline intrusions.

Analcime occurs in the matrix of basic to intermediate volcanic rocks. It may be primary, but rather comes from devitrification of glass. It also appears in the amygdala, bubbles, vesicles of these rocks; it is often associated with zeolites (and calcite, prehnite, axinite, apophyllite, etc.).

Analcime appears as an authigenic mineral in clastic sedimentary rocks, shales (especially lacustrine argillites), silts and sometimes impure sandstones. Some of these rocks contain volcanic material (glass shards, ash, tuffs), but analcime remains a relatively minor component in them. There are rocks with no trace of such a volcanic contribution where analcime appears in a significant proportion (analcimolites) over areas of significant extension and thicknesses (up to 20 m). The fact that analcime is present in such rocks, in the presence of quartz, indicates very little diagenetic evolution. The presence of dolomite and magnesian chlorites suggests a chemical sedimentation.

Analcime + quartz association is characteristic of a very low grade of metamorphism (early zeolite facies). At higher metamorphic grade analcime is replaced by albite.

3.2 MAJOR FERRO-MAGNESIAN MINERALS: MICAS, CHLORITES, AMPHIBOLES, PYROXENES, OLIVINES, SERPENTINES

3.2.1 Micas and related minerals

3.2.1.1 Structure and chemical composition

- Micas are phyllosilicates whose TOT-type sheets are made of two tetrahedral layers sandwiching an octahedral layer (Figure 3.11). The tetrahedral layers consist of a hexagonal pavement of tetrahedra $(SiO_4)^{4-}$ in which each tetrahedron shares three apexes with the neighboring tetrahedra: the chemical composition of such layers is $(Si_4O_{10})^{4-}$. In each sheet, the tetrahedra of the upper tetrahedral layer point downwards, and the ones of the lower tetrahedral sheet point upwards. This pattern

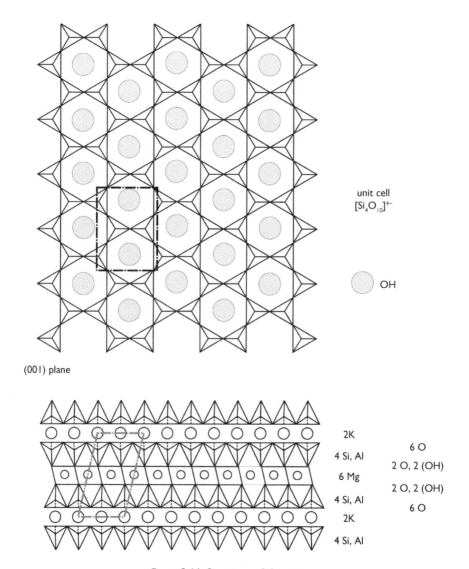

unit cell
$[Si_4O_{10}]^{4-}$

OH

(001) plane

2K

4 Si, Al

6 Mg

4 Si, Al

2K

4 Si, Al

6 O
2 O, 2 (OH)
2 O, 2 (OH)
6 O

Figure 3.11 Structure of the micas.

defines the octahedral sites Y on the intermediate octahedral layer. The filling of the octahedral sites can be 4 Y (dioctahedral mica) or 6 Y cations (trioctahedral micas). The sheets are bound to each other through an interlayer, made of large ions in 12-fold coordination (X site). Anions (OH, F, Cl) are placed at the center of the hexagons of tetrahedral layers. Thus the general formula for mica is:

$$X_2 \, Y_{4-6} \, Z_8 \, O_{20} \, (OH)_4$$

X = K, Na, Ca, (Ba, Rb, Cs...)
Y = octahedral site: Fe, Mg, Al, Ti, Mn, Li, Cr....
Z = tetrahedral site: Si, Al

There are many polymorphs according to the mode of stacking of the different sheets. The most common micas crystallize in the monoclinic (pseudo-hexagonal) system.

Micas are classified according to the nature of the main cation in the X site and filling of the octahedral sheet which may be 2 (dioctahedral micas) or 3 (trioctahedral micas):

X = K, (strictly) common micas, dioctahedral (the type of which is muscovite) and triocathedral (biotite);

X = Na, paragonite (dioctahedral);

X=Ca, brittle micas, diocathedral (margarite) and trioctahedral (clintonite–xanthophyllite).

A type of common dioctahedral mica is muscovite, whose ideal formula is $KAl_2Si_3AlO_{10}(OH)_2$ (half of the unit cell of the above formula). A type of triocathedra mica is biotite, whose simplified ideal formula is $K(Fe,Mg)_3Si_3AlO_{10}(OH)_2$.

The transition from the trioctahedral type to dioctahedral type is made by the muscovitic substitution:

$$3 \, (Mg, Fe) \Leftrightarrow 2 \, Al^{VI} \, \square \quad (\square = \text{vacant site})$$

There is a solvus between dioctahedral micas and trioctahedral micas. The rate of substitution is all the more important the higher the temperature.

Other substitutions occuring in the micas are:

$$Fe^{2+} \Leftrightarrow Mg$$

$Al^{VI} \, Al^{IV} \Leftrightarrow Mg \, Si$ tchermakitic (or phengitic) substitution
$2 \, Al^{VI} \Leftrightarrow (Fe,Mg) \, Ti$ $2 \, (Fe, Mg) \Leftrightarrow Ti \, \square$ $Si \, Al^{VI} \Leftrightarrow Al^{IV} \, Ti$

$$Al^{VI} \Leftrightarrow Fe^{3+}$$
$$K \Leftrightarrow Na$$
$$K \, Si \Leftrightarrow Ca \, A^{IV}$$
$$OH \Leftrightarrow F \Leftrightarrow Cl$$

Calcium enters in very low quantities in the micas. The proportion of Na_2O that may occur in natural muscovites, reaches up to 2 wt%. It is lower in the biotites.

Ferric iron replaces aluminum in the biotites/phlogopites from some volcanic rocks. Fe $\Leftrightarrow$ Mn substitution is extremely limited. Chromium can substitute for octahedral aluminum. Muscovites containing more than 1 wt% Cr_2O_3 are called *fuchsite*.

Among muscovites, tschermakitic substitution $Al^{VI} Al^{IV} \Leftrightarrow (Fe,Mg) Si$ leads from the above ideal formula to *phengite*. Generally low temperature and high pressure muscovites are phengite, the ideal composition of muscovite is reached only for high temperature muscovites (close to the stability limit of muscovite).

$K Al^{IV} \Leftrightarrow \square Si$ substitution leads from muscovite to *sericite* and *illite*. They are usually classified as clay minerals. There is a continuous series between muscovite and illite.

Biotites are classified according to the two mains substitutions $Fe \Leftrightarrow Mg$ and $Al^{VI} Al^{IV} \Leftrightarrow Mg Si$.

Titanium can enter the network of biotite by various substitutions. The type of substitution actually occurring is closely related to geological processes reflected by the evolution of micas. In the presence of a titanium mineral like ilmenite acting as a buffer, the titanium content of biotite increases with temperature, reaching 5 to 6 wt% under conditions of granulite facies. These levels of titanium can be used as a geothermometer (calibrated for pressures of 4–6 kb by Henry et al., 2005). Very little titanium can enter in the network of muscovite.

3.2.1.2   Alterations of micas

Biotite is very alterable mineral and is commonly altered into chlorite at low temperatures, in particular, in surface conditions. The titanium content in the lattice then recrystallized as grains or fine needles of rutile, sometimes showing the sagenite twinning.

Epidote and prehnite occur less commonly in elongated lenses in the cleavages of biotite.

Muscovite is a hardly alterable mineral; it is common as clasts in clastic sedimentary rocks.

3.2.1.3   Stability of micas

Illite is stable in the sedimentary environment and its gradual recrystallization produces a mica of muscovite–phengite type. The proportion of phengitic substitution in muscovite decreases as temperature increases, until a composition of near ideal muscovite at the stability limit of muscovite.

The stability of muscovite is limited to high temperatures by the discontinous reaction (Figure 3.11):

Muscovite + quartz => K-feldspar + sillimanite + water

Phengites replace potassium feldspar at high pressure. In the presence of potassium feldspar, phengitic substitution rates in muscovite can be used as a barometer.

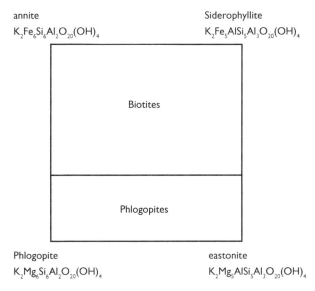

annite
$K_2Fe_6Si_6Al_2O_{20}(OH)_4$

Siderophyllite
$K_2Fe_5AlSi_5Al_3O_{20}(OH)_4$

Biotites

Phlogopites

Phlogopite
$K_2Mg_6Si_6Al_2O_{20}(OH)_4$

eastonite
$K_2Mg_5AlSi_5Al_3O_{20}(OH)_4$

Figure 3.12 Biotites.

Biotite is not stable at low temperatures but is replaced by muscovite + chlorite associations or stilpnomelane. Biotite appears towards the middle part of the greenschist facies, that is at temperatures of about 400–500°C.

Towards higher temperatures biotite becomes unstable and is destabilized by reactions similar the reactions to the disappearance of muscovite:

Biotite + quartz => orthopyroxene + feldspar + water
Aluminous biotite + quartz => garnet + orthoclase + water

These are continuous reactions as the minerals involved contain both iron and magnesium. The ferriferous terms are destabilized at temperatures lower than the magnesian terms.

Stability of iron-bearing micas also depends on oxygen fugacity. Annite is stable to higher temperatures as oxygen fugacity is lower. Phlogopite is stable up to temperatures of about 900°C at low water pressure and at about 1100°C at water pressures of 4–5 kb.

3.2.1.4 Occurrences of micas

Igneous rocks

In Bowen's reaction series, micas crystallize after pyroxene and amphibole and more or less simultaneously with acid plagioclase and potassium feldspar. Biotite often begin to crystallize in the same time as amphibole and plagioclase, of intermediate composition. Muscovite appears at the end of crystallization with quartz and feldspar.

The biotite is thus a very common constituent of intermediate and acidic igneous rocks, volcanic and plutonic, silica saturated or not, alumina saturated or not. It is particularly characteristic of the intermediate meta-aluminous rocks. Its composition depends, of course, on the composition of the host rock, in which is often the major ferro-magnesian mineral.

In a series diorite – granodiorite – granite – pegmatite, biotite varies from more magnesian and titanium-richer terms in ibasic, to intermediate rocks, to iron-richer terms in acidic rocks. Alumina contents also increases: in the case of fractional crystallization under the influence of minerals such as pyroxene, amphibole (or less aluminous biotites), differentiated magmas are enriched in aluminum, which leads to the crystallization in the final terms of aluminous biotite and muscovite.

The composition of biotite also reflects the type of magmatic series.

Compare, for instance, primary magmatic biotites in various magmatic suites:

Sain.-Arnac granite (Pyrenees Orientales, France): calc-alkaline granite massif emplaced at an higher level of the crust (roof in the chlorite zone) (Touil, 1994):

- Ansignan charnockitic granite (Pyrenees Orientales, France): granulite facies (Touil, 1994).
- Laouzas peraluminous magnesio-potassic granite (Montagne Noire, France) emplaced in K-feldspar + sillimanite zone (Demange, 1982 and unpublished data).
- Sierra dos Orgaõs garnet granite tholeiitic (Rio de Janeiro, Brazil) lately (post-metamorphism) emplaced in rocks in amphibolite facies at the base and of greenschist facies rocks at the roof (Demange and Machado, 1998).

Variations in biotite composition in each series is compatible with the model of fractional crystallization: decreases of titanium content, increases of Fe/Fe + Mg ratio with differentiation. The comparison between these sequences shows:

- The influence of the level of emplacement: titanium content is increased as the level of emplacement is deeper, that is the temperature of the host rocks is higher. In Ansigan charnockitic granite alone, the only stable biotites, stable at the magmatic stage, are magnesian biotites;
- the magnesian (Laouzas) or ferriferous (Orgaõs) character of the series; note that the presence of iron-rich garnet in the Orgaõs series is more the consequence of the high level of iron in this series than any peraluminous character;

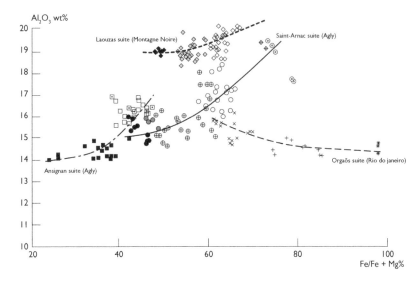

Figure 3.13 Composition of the biotites of granitoids: $Al_2O_3/X_{Fe.}$ ($X_{Fe} = 100\ Fe/(Fe + Mg)$) diagram.

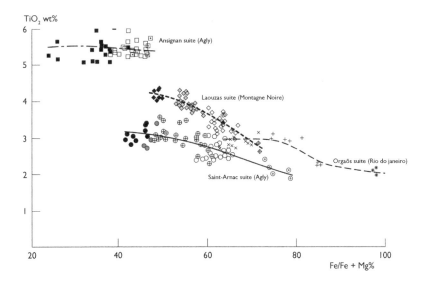

Figure 3.14 Composition of the biotites of granitoids: TiO_2/X_{Fe} ($X_{Fe} = 100\ Fe/(Fe + Mg)$) diagram.

Saint-Arnac suite (Agly): diorites- quartz-diorites ●, granodiorites ⊕, pseudoporphyritic granite ○, leucocratic granite ⊙ Ansignan suite (Agly): norites ■, charnockitic granite □, white garnet granite ▣ Laouzas suite (Montagne Noire): dark cordierite granodiorite ◆, dark cordierite granite ◇, clear biotite granite ◇ hololeucocratic granite with cordierite nodules ⊕ Orgãos suite (Rio de Janeiro, Brésil): amphibole tonalites – granodiorites x, biotite granites +, pegmatites ✱.

- biotites in the common calc-alkaline series (granites and Saint-Arnac and Ansignan granites) and iron-rich series (Orgaõs) have medium Al_2O_3 content 14 to 18 wt%; this content increases greatly in differentiated terms in the Saint-Arnac and Ansignan series. The Laouzas peraluminous biotite granite have Al_2O_3 contents of about 19–22%. For comparison, the Al_2O_3 content of biotite in nepheline syenites are about 12–13%.

In charnockites that are granitoids that crystallized in the condition of the granulite facies, biotites appear only in late phases (crystallized at higher water pressures) or as a secondary mineral that develops on the (ortho) pyroxenes.

Biotite/phlogopite is less common in basic rocks. Indeed the limit of stability of phlogopite is close to the temperature of crystallization of basalt. However biotites are stabilized in such rocks by higher water pressure, the presence of fluorine, alkaline and especially the potassic character of the magma: so biotite is present in basanites and alkali gabbros (theralites, etc.). Biotite is also a major constituent of lamprophyres and kimberlites.

The limit of stability of muscovite is close to the temperature of crystallization of granite. It is much less common in igneous rocks than biotite: a comparison of stability curves of muscovite and the melting of granite (Figure 3.11) shows that muscovite can crystallize only under water pressures of at least 1.5 kb. This stability range is increased by the presence of fluorine. Muscovite therefore only appears in evolved plutonic rocks, granites and pegmatites. Muscovite as a primary magmatic mineral, is an indication of the peraluminous character of these rocks. But in the plutonic rocks, muscovite is often a secondary mineral formed by alteration of feldspar (leaching of alkalis) or of aluminous minerals such as cordierite.

Metamorphic rocks

The field of stability of muscovite/phengite covers metamorphic facies of low temperature until the middle part of the amphibolite facies. That of biotite covers the upper greenschist facies and amphibolite facies. Biotite disappears in the granulite facies.

Metamorphism of low to medium pressure (and contact metamorphism) transforms the more or less clay-rich sedimentary rocks into schists rich in biotite and muscovite. In low-grade metamorphism (epizone), sedimentary illites recrystallize into sericite and muscovite. The biotite isograd marks the medium grade metamorphism (mesozone). Aluminum-rich rocks (metapelites, meta-shales) contain more aluminous minerals (silicates of alumina, chloritoid, staurolite, cordierite, garnet, etc.). The isograd of the disappearance of muscovite can be taken as the limit of high grade metamorphism (catazone). In high grade metamorphism, metapelites are transformed into aluminous gneisses with biotite, potassium feldspar, plagioclase, sillimanite, possibly cordierite and/or garnet. Micas in such rocks are rich in alumina: Al_2O_3 contents of biotite are about 19–21 wt%.

The contents of Na_2O and TiO_2 increase with metamorphic grade: for example, in Montagne Noire, TiO_2 content of biotite increases from 1.6 wt% at the biotite isograd to 2.7–2.9 wt% in the sillimanite + muscovite zone (the content of muscovites increases from 0.3–0.7 wt%). In the same interval the contents of Na_2O in muscovites grow from 0.4–0.5 to 1–1.1 wt% (0.10–0.15 to 0.25–0.30% in biotites); this is without any paragonite in the paragenesis.

Iron-magnesium ratios vary according to more complex laws and depend on the parageneses, the metamorphic grade and the type of metamorphism. This last point will be developed in the chapter on aluminous minerals.

The composition of the biotite in orthogneiss of intermediate to acidic reflects the compositions of biotites of the initial rocks and are clearly distinguishable from biotites from metapelites particularly by their contents of aluminum and titanium. It is the same for the biotite ortho-amphibolites derived from basic igneous rocks; the common paragenesis of these rocks is plagioclase – amphibole (hornblende in amphibolite facies) – biotite (± garnet).

Phlogopite is a common constituent of impure marbles and calc-silicate-gneisses. Ordinary biotite may exist in association with the amphibole and plagioclase in some (para-) calc-silicate-gneisses – poor in calcium. But if the calcium content increases biotite is replaced by amphibole.

Biotite is not stable in high pressure *metamorphisme. It is then* replaced by phengite + chlorite and/or stilpnomelane. Phengite also replaces potassium feldspar, particularly in orthogneisses.

Metasomatic rocks

Muscovite is a secondary mineral developing on potassium feldspar by the leaching of potassium: its habit is large mottled crystals ("skelettal" or "spongy" muscovite). Plagioclase is altered to sericite by Na–Ca leaching and the addition of potassium. Alumina silicates and aluminum silicates (cordierite, staurolite) are also altered into muscovite, especially during stages of retrograde metamorphism.

Further leaching of alkali led to *greisen*, rocks formed mainly of quartz + muscovite. Greisens often contain beryl, topaz, tourmaline, fluorite, cassiterite, wolframite, etc. The greisens form mainly on granitoids and pegmatites. They can also develop on more varied rocks as far as they are aluminous enough. Muscovitites are related rocks, in which quartz is completely leached.

Potassic alteration transforms amphibole into biotite. Some mafic and ultramafic rocks can be transformed into biotitites in contact with granitic magmas and pegmatites. Biotitites can also form in halos in hydrothermal ore deposits: in the gold Salsigne deposit (Aude, France), biotitites develop from silts and pelites around the orebodies on decimeter to meter thicknesses (Demange et al. 2006).

3.2.1.5 Lithium-bearing micas

The composition of lithium-bearing micas are deduced from those of the common micas by substitutions:

$Al^{VI} \square \Leftrightarrow 2\ Li$
$Al^{VI} \square \Leftrightarrow Li\ M^{2+}$
$2\ R^{2+} \Leftrightarrow Al^{VI}\ Li$

where R^{2+} refers to a divalent cation and $\square$ a vacant site

Introduction of lithium in the lattice is accompanied by the introduction of fluoride. There are miscibility gaps, yet insufficiently explored, and discontinuities in the crystallization, the results of which are discontinuous zonations observable under the microscope (cross polarized light).

The general term lepidolite describes purely aluminous lithium-bearing micas intermediate between di- and tri-octahedral micas. Zinnwaldite are trioctahedral micas; they are practically devoid of magnesium.

Lithium-bearing micas are rare minerals that occur in lithic pegmatites where they are associated with spodumene, amblygonite ($LiAlPO_4$ (F, OH)), petalite, tourmaline, topaz, beryl, albite, quartz, cassiterite, fluorite. They are also found in hydrothermal veins (and greisen).

3.2.1.6 Paragonite

Paragonite $Na_2Al_4Si_6Al_2O_{20}(OH)_4$ is the sodic equivalent of muscovite, which is indistinguishable under the microscope. There is a solvus between muscovite and paragonite and the proportion of potassium entering paragonite, as with sodium entering muscovite, increases with temperature. The natural paragonite may contain up to 4 wt% of K_2O.

Paragonite is a metamorphic mineral. Its occurrences are similar to those of muscovite: meta-pelites from low to medium temperature (with kyanite, staurolite, garnet) from low to high pressure. It occurs particularly in blueschist and eclogite facies.

3.2.1.7 Brittle micas

Margarite $Ca_2Al_4Si_6Al_2O_{20}(OH)_4$ is a dioctahedral brittle mica, calciuc equivalent to muscovite.

Clintonite and xanthophyllite $Ca_2(Mg, Al)_6Si_{1,5}Al_{5,5}O_{20}(OH)_4$ are trioctahedral micas. They are the equivalent of phlogopite. The iron only substitutes very little to magnesium. These two minerals are distinguished by their optical properties: α is perpendicular to (010) in clintonite, parallel in xanthophyllite.

Brittle micas are rare minerals.

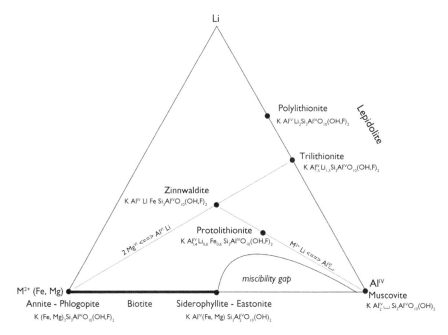

Figure 3.15 Lithium-bearing micas.

Margarite occurs mainly in emery deposits (metamorphosed bauxites) with corundum, diaspore. Also appears in iron-poor (calcic) peraluminous schists with kyanite, or andalusite. The sample featured on the CD from Ovala (Gabon), is a rock of metasomatic origin, where margarite is associated with kyanite and anorthite (Demange, 1976).

Clintonite and xanthophyllite are minerals from calcic and magnesian schists where they are associated with chlorite, from metamorphic dolostones and skarns.

3.2.1.8 *Stilpnomelane*

Stilpnomelane is a phyllosilicate with a structure similar to that of micas: with sheets made of two tetrahedral layers separated by an octahedral layer:

- the octahedral layer is complete,
- the tetrahedral layers are incomplete and are formed by regularly disposed "islands" of 7 rings of $(SiO_4)^{4-}$ tetrahedra,
- the interlayer (K, with small amounts of Na and Ca) is incomplete.

Several formulas have been proposed:

$(K, Na, Ca)_{0.6} (Fe^{3+}, Fe^{2+}, Mg, Mn)_6 Si_8Al (O, OH)_{27} \cdot 2\text{–}4\ H_2O$

$(K, Na, Ca)_{0.7} (Fe^{3+}, Fe^{2+}, Mg, Mn, Al)_8 (Si,Al)_{12} (O, OH)_{27} \cdot n\ H_2O$

Stilpnomelane is essentially an iron-rich mineral. The iron may be in the ferrous state (ferrostilpnomelane) or ferric (ferristilpnomelane). The transition from one to another occurs by the substitution:

$Fe^{2+} (OH)^- \Leftrightarrow Fe^{3+}\ O^{2-}$

Stilpnomelane is a rather common mineral in metamorphic metapelites (and basic rocks) of greenschist facies (with chlorite, muscovite, albite, eventually epidote, actinolite) and blueschist facies (with phengite, garnet, glaucophane).

It is also a major constituent of iron ore of banded iron formation (BIF) type with other iron-bearing minerals: magnetite, minnesotaïte (ferroan equivalent of talc) grunerite (orthorhombic amphibole) chamosite, greenalite (oxychlorites).

3.2.1.9   Talc

Talc is a phyllosilicate close to the micas, but its occurrences are different. It will be treated with the magnesian minerals (2.4.1d).

3.2.1.10   Zussmanite – Howieite – Deerite

Zussmanite, $K(Fe^{2+}, Mg, Mn)_{13}AlSi_{17}O_{42}(OH)_{14}$ (trigonal) is a phyllosilicate, somewhat similar to trioctahedral micas: a continous octahedral layer containing iron, with tetrahedral (Si,Al) attached on both sides, linked by potassium atoms and tree-member rings of tetrahedra.

Howieite, $Na(Fe^{2+}, Mn^{2+})_{10}(Fe^{3+}, Al)Si_{12}O_{31}(OH)_3$ (triclinic) and deerite $((Fe^{2+}, Mn^{2+})(Fe^{3+}, Al)_3Si_6O_{20}(OH)_5$ (monoclinic) are inosilicates, with 4 pediodic single chains. In howeite ribbons of octahedral layer, similar to the ones of micas, are sandwiched between two $[Si_6O_{17}]$ chains.

Experimental data (Dempsey, 1981) indicate that zussmanite is stable between 10–30 kb, at temperatures up to 600°C. The lower pressure range of deerite is estimated to 4 kb at 200°C, 6 kb at 300°C (Muir Wood, 1972).

These are rare minerals of metamorphism of iron-rich formations in blue schist facies. They are commonly associated, and occur with such iron-rich (and manganese-rich) minerals as stilpnomelane, spessatine, riebeckite, aegyrine, grunerite.

3.2.2   Chlorites

3.2.2.1   Structure and chemical composition

Chlorites are phyllosilicates formed from the superposition of talc-type layers similar to that of mica (two tetrahedral layers surrounding an octahedral layer $Y_6Z_8O_{20}(OH)_4$) and brucite-type layers formed by an octahedral layer $Y_6(OH)_{12}$.(Figure 3.40).

The general formula of chlorite is:

$$Y_{12}\ Z_8\ O_{20}\ (OH)_{16}$$
$$Y = Fe^{2+}, Mg, Mn, Al, Fe^{3+}, Cr^{3+}$$
$$Z = Si, Al$$

Chlorites vary mostly by the substitutions:

$$Fe^{2+} \Leftrightarrow Mg$$
$$Al^{VI}\ Al^{IV} \Leftrightarrow Mg\ Si$$
$$Al^{VI} < = > Fe^{3+}$$

The non-aluminous magnesian end member $Mg_{12}Si_8O_{20}(OH)_8$, is not a chlorite but antigorite (a serpentine group mineral). Manganese and chromium are generally very minor constituents.

Chlorites are divided into (ordinary) chlorites and oxychlorites (or septochlorites) by the proportion of ferric iron. By convention, the boundary between chlorite and oxychlorites is fixed to a content of 4 wt% of Fe_2O_3.

Each group is further divided into different terms according to iron/magnesium and silicon/aluminum ratio (Hey's diagram, 1954, see Figure 3.16) This nomenclature is not very useful.

Sudoite is a dioctahedral (thus rich in aluminum) magnesian chlorite, of formula $(Mg_4Al_6)(Si_6Al_2)O_{20}(OH)_8$. It is stable only at (very) low temperature (anchimetamorphism, beginning of epizone).

3.2.2.2   Occurrences of chlorites

Chlorites are minerals rich in water and are not stable at high temperatures. They occur in sedimentary rocks and low to medium grade metamorphism. Chlorites are not of primary minerals of igneous rocks.

Sedimentary rocks

Chlorites are important constituents of the sedimentary rocks as clastic or authigenic minerals. They can also come from the diagenetic recrystallization or anchimetamorphism of clays of the smectite group.

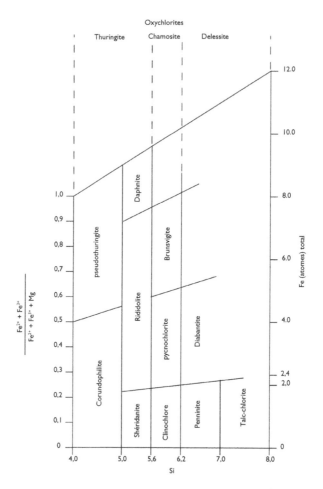

Figure 3.16 Nomenclature of the chlorites (after Hey, 1954).

Oxychlorites are constituents of sedimentary iron ores: chamosite (aluminous) in oolitic hematitic ore type (Clinton type, Anjou – Normandy district in France) or limonitic ore type ("minette" type, Lorraine, France), greenalite (non-aluminous) and chamosite in banded iron formations (BIF).

Metamorphic rocks

Chlorite is an important mineral in rocks of sedimentary origin and of low to medium grade metamorphism. In low to medium pressure metamorphism, prograde reactions of appearance of the various index minerals gradually restrict the field of composition of the chlorite: appearance of biotite, then of aluminous magnesian and/or ferriferrous silicates (cordierite, garnet, staurolite) and finally alumina silicates. Usually chlorite disappears in the andalusite or kyanite zone (see the section on aluminous minerals, 3.3).

Chlorite is a common mineral in the blueschist facies in association with albite, glaucophane, phengite, etc. In white schist facies, the association talc – kyanite replaces magnesian chlorite.

In the basic rocks of igneous origin, the association chlorite – albite – epidote – actinolite is characteristic of the greenschist facies. This association is replaced by the paragenesis hornblende – epidote – albite/oligoclase in the epidote – amphibolite facies and by hornblende – calcic plagioclase in the amphibolite facies.

Steatite (soapstone) is a rock made of talc, chlorite, actinolite, etc., and of low to medium grade metamorphism of ultramafic rocks (for example, dunite, harzburgite).

Chlorite is also present in the intermediate and acidic rocks where it replaces the primary biotite and amphibole.

Secondary mineral

Chlorite is a common alteration mineral of ferromagnesian minerals (biotite, amphibole, pyroxene, olivine, garnet, cordierite, etc.) It may be supergene alteration, but most often an alteration associated with hydrothermal fluids circulation or deuteric alteration the end of crystallization of igneous rocks, or even a retrograde mineral in the lowering of temperature in metamorphism.

Biotites are altered frequently into a green chlorite, with Prussian blue interference colors. Titanium contained in the biotite recrystallizes in rutile in grains, often showing the habit of sagenite (60° twinned needles).

Cordierite is typically altered into a yellowish isotropic chlorite.

Metasomatic rocks

The rocks, once called "spilites" (a term now obsolete), show occurrences and textures of basaltic lava, but they are formed of albite and chlorite (± calcite–actinolite–epidote). Comparison to similar lavas shows a leaching of calcium and sodium enrichment. They are commonly attributed to oceanic hydrothermal metamorphism, but they may also occur in a continental environment.

Propylitization (propylitic alteration) is an alteration developed in mafic to intermediate (igneous) rocks around hydrothermal sulfide deposits: it transforms the primary rock into an association epidote–albite–chlorite–pyrite (sometimes with hematite and magnetite).

3.2.3 Amphiboles

3.2.3.1 Structure and chemical composition

Amphiboles are inosilicates where tetrahedra $(SiO_4)^{4-}$ are organized in double chains $(Si_4O_{11})^{6-}$. The unit cell includes the equivalent of two double chains and comprises 8 tetrahedral sites Z (Figure 3.17). These chains are linked by 5 octahedral Y sites (6-fold coordination) and two X sites which have 6- or 8-fold coordination. There is also a large A site of 12-fold coordination, which may be full or empty. OH ions are at the center of the hexagons that form the chain in the plane of the apexes of the tetrahedra.

Thus general formula of amphiboles is:

$$A_{0-1} X_2 Y_5 Z_8 O_{22} (OH)_2$$

Z = site tetrahedral site: Si, Al
Y = site octahedral site: Mg, Fe^{2+}, Fe^{3+}, Al, Ti, Mn, (Cr, Li)
X = site of 8- or 6-fold coordination: Ca, Na, K, Fe, Mg (Mn, Li)
A = site of 12-fold coordination 12: Na, K

The word αμφιβολοσ means misleading: under the very similar external aspects, amphiboles shows a wide range of chemical composition. The precise determination of amphibole requires microprobe. The composition of amphiboles (other than ferromagnesian amphiboles) is deduced from the composition of tremolite □ $Ca_2Mg_5Si_8O_{22}(OH)_2$ (the A site is empty)by the following substitutions:

$R^{2+} Si \Leftrightarrow R^{3+} Al^{VI}$
$Ca_x R^{2+} \Leftrightarrow Na_x R^{3+}$
$Ca_x \square \Leftrightarrow Na_x Na_A$
$Si \square \Leftrightarrow Al^{IV} Na_A$
$Fe \Leftrightarrow Mg \Leftrightarrow Mn \qquad Al^{VI} \Leftrightarrow R^{3+}$
$Na \Leftrightarrow K$
$OH \Leftrightarrow F \Leftrightarrow Cl$
$2 Al^{VI} \Leftrightarrow (Fe,Mg) Ti \qquad 2 (Fe, Mg) \Leftrightarrow Ti \qquad Si AlVI \Leftrightarrow Al^{IV} Ti$
$2 R^{2+} \Leftrightarrow Al^{VI} Li \qquad Al^{VI} \square \Leftrightarrow R^{2+} Li$

where R^{2+} is a divalent cation, R^{3+} a trivalent cation, □ a vacant site

These (independent) substitutions form a 15 dimension space. There are certainly miscibility gaps in this space but not all of them are investigated here.

3.2.3.2 Classification of the amphiboles (Leake, 1978)

Amphiboles are classified according to the filling of the X site: Fe, Mg, Ca or Na (Table 3.1 lists the end members).

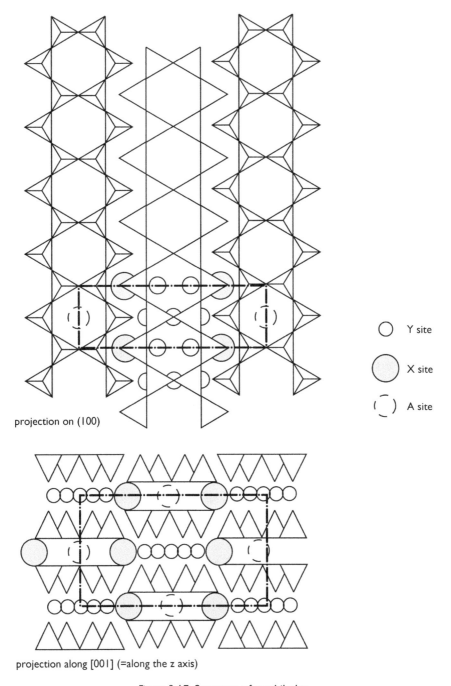

projection on (100)

○ Y site

◯ X site

(⌒) A site

projection along [001] (=along the z axis)

Figure 3.17 Structure of amphiboles.

Table 3.1 Classification of the amphiboles.

Amphiboles	A_{0-1}	X_2	Y_5	Z_8
ferro- magnesian amphiboles				
anthophyllite – gedrite series				
anthophyllite		Mg_2	Mg_5	Si_8
gedrite		$(Mg, Fe)_2$	$(Mg, Fe)_3 Al_2$	$Si_6 Al_2$
cummingtonite – gedrite serie		$(Mg, Fe)_2$	$(Mg, Fe)_5$	Si_8
calcic amphiboles ($Na^x < 0,67$)				
non aluminous				
tremolite – actinolite –				
ferroactinolite series		Ca_2	$(Mg, Fe)_5$	Si_8
aluminous				
horblendes				
tschermakite –				
ferrotschermakite		Ca_2	$(Mg, Fe)_3 Al_2$	$Si_6 Al_2$
edenite – ferroedenite	Na	Ca_2	$(Mg, Fe)_5$	$Si_7 Al$
pargasite –				
ferrohastingsite	Na	Ca_2	$(Mg, Fe)_4 Al$	$Si_6 Al_2$
brown hornblendes				
basaltic hornblende	$(Ca, Na)_{2-3}$		$(Fe^{2+}, Mg)_{3-2} (Fe^{3+}, Al)_{2-3}$	$Si_6 Al_2$
kaersutite	(Na, K)	Ca_2	$(Mg, Fe)_4 Ti$	$Si_6 Al_2$
barkevikite	(Na, K)	Ca_2	$(Fe^{2+}, Mg, Fe^{3+}, Mn)_5$	$Si_{6,5} Al_{1,5}$
sodi-calcic amphiboles	**($0,67 > Na^x < 1,34$)**			
barroisite				
– ferrobarroisite series		Na Ca	$(Mg, Fe)_3 Al\ Fe^{3+}$	$Si_7 Al$
série richterite – katophorite				
richterite				
(soda-trémolite) –				
ferrorichterite	Na	Na Ca	$(Mg, Fe)_5$	Si_8
magnesiokatophorite –				
katophorite	Na	Na Ca	$(Mg, Fe)_4 (Al, Fe^{3+})$	$Si_7 Al$
sodic amphiboles	**($Na^x > 1,34$)**			
glaucophane – riebeckite				
series (crossites)				
glaucophane		Na_2	$Mg_3 Al_2$	Si_8
riebeckite		Na_2	$Fe^{2+}_3 Fe^{3+}_2$	Si_8
riebeckite – arfvedsonite –				
(eckermannite) series				
arfvedsonite	Na	Na_2	$Fe^{2+}_4 Fe^{3+}$	Si_8
eckermannite (the				
pure end member				
is non known)	Na	Na_2	$Mg_4 Al$	Si_8

Ferro-magnesian amphiboles

–   anthophyllite Mg_7Si_8 – gedrite $Mg_5Al_2Si_6$ series: orthorhombic amphiboles (all the other amphiboles are monoclinic).
–   cummingtonite $(Mg, Fe)_7Si_8$ – grunerite $(Fe, Mg)_7Si_8$ series.

They are quite rare minerals which occur primarily or exclusively (anthophyllite-gedrite series) in high grade regional metamorphism of mafic and ultramafic rocks, and for grunerite in metamorphosed iron deposits.

Calcic amphiboles

Calcic amphiboles are divided into three groups:

•   non-aluminous calcic amphiboles form a continuous series tremolite (Mg) – actinolite Ca_2 (Fe, Mg) 5 Si_8 – ferroactinolite (Fe);
•   aluminous calcic amphiboles or hornblendes. These are the most common amphibole, both common in igneous and metamorphic rocks. They are generally described as a proportion of the end members: edenite – ferroedenite, tschermakite – ferrotschermakite, pargasite – hastingsite – ferrohastingsite. There is a gap of solubility at low temperatures between the tremolite–ferroactinolite series and hornblende; beyond 720°C the solubility is complete;
•   brown hornblendes are distinguished by their high iron content (particularly ferric iron) (basaltic hornblende, barkevicite) and/or titanium (kaersutite) and their high sodium contain. Under the microscope they have a brown, reddish brown, yellowish brown color. These are minerals of volcanic rocks (and igneous rocks kaersutite, barkevicite) of basic to intermediate composition. Kaersutite and barkevicite are characteristic of alkaline rocks.

Sodi-calcic amphiboles

Sodic-calcic amphiboles have NaCa position X. They are classified into two groups:

•   the barroisite (– ferrobarroisites) are aluminous amphiboles, whose A site is empty. These minerals are mainly found in high-pressure metamorphism.
•   presence of sodium in A site places the amphibole of the richterite–ferrorichterite–katophorite group in a position intermediate to sodic amphiboles. Ferrorichterite–katophorite contain ferric iron whose presence induces yellow brown (ferrorichterite) to reddish brown colors (katophorite) in thin section. Katophorite is a mineral of basic alkaline rocks. Richterite is found in metamorphic marbles and skarns.

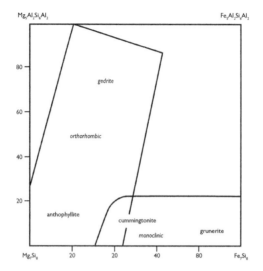

Figure 3.18 Ferro-magnesian amphiboles (after Tröger, 1961).

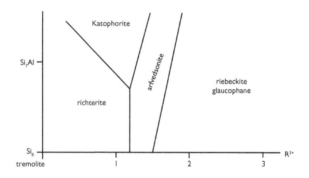

Figure 3.19 Sodi-calcic and sodic amphiboles.

Sodic amphiboles

Sodic amphiboles, whose X site is occupied by sodium, form two groups:

- glaucophane–riebeckite series, whose intermediate terms are called crossites, are high pressure metamorphic minerals. Glaucophane is equivalent to the association chlorite + albite;
- riebeckite–arfvedsonite–eckermannite series. The pure eckermannite end member does not exist in the nature, natural eckermannite still containing some calcium. These are mineral acidic or intermediate alkaline igneous rocks, saturated in silica or not.

3.2.3.3 Stability of amphiboles

The stability field of the amphibole is limited towards the low temperatures – where they are replaced by chlorite, serpentine, with or without carbonates – and towards the high temperatures where they are replaced by pyroxenes.

Amphibole stability depends both on their chemical composition, and thus the chemical composition of the rock, and on physical parameters:

- like all other ferromagnesian minerals, magnesian amphiboles are stable at higher temperatures than the amphibole containing iron;
- high water pressures stabilize amphibole rather than pyroxene;
- oxygen fugacity is important in that iron can be in ferric or ferrous state.

Generally the upper limits of stability are the order of 500–800C at 0.5 kb and 550–1000°C at 2 kb. These are the conditions of plutonism and metamorphism of medium grade.

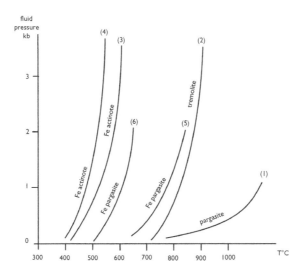

Figure 3.20 Stability of amphiboles (after Boyd, 1959, Gilbert, 1966 and Jenkins and Bozhilov, 2003).

1	pargasite = > aluminous diopside + forsterite + nepheline + spinel + H_2O
2	tremolite = > diopside + enstatite + quartz + H_2O
3	ferroactinote = > hedenbergite + fayalite + quartz + magnetite + H_2O
	f_{O2} fixed by the iron = wustite buffer
4	same reaction f_{O2} fixed by the fayalite = magnetite + quartz buffer
5	ferropargasite = > hedenbergite + fayalite + magnetite + plagioclase + nepheline + H_2O
	f_{O2} fixed by the iron = wustite buffer
6	same reaction f_{O2} fixed by the fayalite = magnetite + quartz buffer

3.2.3.4 Occurrences of amphiboles

By the wide range of their chemical compositions, amphiboles have extremely varied deposits, reflecting both the chemical composition of the rock and its conditions formation. In a group of amphiboles belonging to a complex of related rocks (magmatic series, metamorphic suite), changes in chemical composition and the laws of substitution between them, reflect very precisely the geological evolution of these rocks. It is worthwhile to study them even if the experimental data available are still insufficient to interpret these geochemical results in terms of precise physical. (emperature, pressures, etc.) estimations.

Metamorphic rocks

The metamorphic facies (recalled in the introduction §1.3.1 and Figure 1.9) are largely defined by the presence, or absence, of amphiboles and their nature.

In *low to medium pressure metamorphism*, amphiboles are generally calcic amphiboles and appear in rocks derived from basic to intermediate igneous rocks and in carbonate rocks.

In *rocks deriving from igneous rocks*, in greenschist facies, amphibole is a non-aluminous calcic amphibole of the tremolite–actinolite series (with chlorite, epidote, albite, etc.). In the epidote-amphibolite facies, amphibole is a hornblende associated with epidote and sodic plagioclase (An < 20). Amphibolite facies itself is defined by the critical association hornblende and calcic plagioclase. With increasing metamorphism the composition of amphibole varies: amphibole becomes progressively richer in sodium, changing from terms richer in tschermakitic component to terms richer in edenitic component. Amphiboles of low pressure are less rich in pargasitic component than the ones under pressure. These compositional changes induce variations of the optical properties: green to bluish green at low temperature, amphibole becomes brownish green and then brown green at higher temperatures.

Cummingtonite appears in the upper amphibolite facies where it results from destabilization of the iron-rich component of hornblende into cummingtonite + calcic plagioclase. Cummingtonite coexists with a hornblende that is then more magnesian and less aluminous (edenitic type). At high temperatures amphiboles (hornblende and cummingtonite) disappear and are replaced by the association orthopyroxene + calcic plagioclase characteristic of granulite facies.

In *metamorphic carbonate rocks*, the chemical compositions of amphiboles also vary depending on the intensity of metamorphism. The variations are larger than in the igneous rocks but they are less clearly correlated to the intensity of metamorphism due to the initial variability of the rocks (calcitic or dolomitic marble, proportions of clay, silica, etc.). Amphiboles of the tremolite – actinolite series appear from the biotite isograd. Tremolite is stable at temperatures higher than those of the greenschist facies in dolomitic marbles. In the amphibolite facies, amphiboles of the other carbonate rocks are hornblende often rich in pargasitic constituent and/or richterites.

Some *rocks of particular chemical composition* contain uncommon amphiboles. Grunerite is restricted to metamorphic iron ores or sediments rich in iron (Collobrieres Var; Pierrefitte Hautes Pyrenees, France). Riebeckite also appears in sediments rich in iron (Saint-Veran, Hautes Alpes, France). Anthophyllite appears in highly magnesian rocks (ultrabasic or ultramafic rocks initially more or less rich in olivine and orthopyroxene). It has a very limited range of stability: at temperatures lower than about 700°C it is replaced by chlorite and serpentine; at temperatures higher than 800°C, it is replaced by orthopyroxene. Anthophyllite is also described in rocks of sedimentary origin, probably having undergone metasomatic phenomena: cordierite – anthophyllite gneiss (famous examples in Finland), contact metamorphism of the Pyrenean lherzolites.

In high-pressure metamorphism, blueschist facies is defined by the presence of amphibole of the glaucophane – riebeckite series (crossite). Crossites have a compositions similar to the one of the association chlorite + albite. Thus, these minerals appear in rocks of more varied composition in low to medium pressure metamorphism: meta-basites, but also meta-greywackes and meta-pelites.

Igneous rocks

Besides tschermackite which is peraluminous, amphiboles are meta-aluminous (hornblendes) or alkaline (arfvedsonite–riebeckite series, katophorite) minerals.

Common hornblende is an important and frequent constituent of acidic to intermediate *volcanic rocks*, saturated in silica or not. It is less abundant in basic rocks. Crystallization of amphibole instead of pyroxene depends on water pressure. Crystallization of one or other of these minerals influences the subsequent evolution of the magma. Hornblendes of volcanic rocks usually belong to the pargasite–ferrohastingsite series.

Basaltic hornblende is common in mafic to intermediate rocks, in saturated silica or not. It is generally accepted that these minerals derive from a primary common hornblende by oxidation during or after the eruption.

Cummingtonite is a rare constituent of some dacites.

In alkaline series, katophorite and barkevicite appear in mafic to intermediate rocks, kaersutite (Ti) in intermediate to acidic rocks, while the differentiated rocks show an amphibole of riebeckite (Fe^{3+}) type. Ferrorichterite occur in sodium-rich alkaline rocks (syenites and granites). Richterite also occurs in carbonatites and meteorites.

Hornblende is a very common constituent of basic (or even ultrabasic), intermediate and more rarely, acidic *igneous rocks*. It is more common than in the corresponding volcanic rocks because in igneous rocks the evolution of the magma occurs under water pressures higher than in volcanic rocks. Amphibole reflects the composition of magmatic evolution: magnesian hornblende in gabbros, ferrohastingsite in granites and nepheline syenite.

Amphiboles of the arfvedsonite–riebeckite series appear in the acidic alkaline rocks, under-saturated or not in silica. Barkevicite is a rare constituent of alkaline basic to intermediate rocks under-saturated in silica like theralite (nepheline gabbros) essexite (nepheline monzogabbros), nepheline syenite or highly under-saturated rocks like jacupirangites (rocks made of titaniferous augite, magnetite and minor nepheline).

Amphibole as a secondary mineral

Ouralite is a common alteration by hydratation of pyroxene into amphibole (actinolite or hornblende).

Kelyphite is an alteration of a pyrope-rich garnet into amphibole + plagioclase, as a retromophic reaction from eclogite facies to amphibolite facies.

3.2.4 Pyroxenes

Pyroxenes are most important ferromagnesian minerals: they are major constituents of many igneous rocks and also occur in varied metamorphic rocks.

3.2.4.1 *Structure and chemical composition*

Pyroxenes are single chain inosilicates $[(SiO_3)^{2-}]_n$ (Figure 3.21). The chains are linked by chains of octahedra (6-fold coordinated Y sites) and 8-fold coordinated X sites. The general formula of pyroxene is thus:

$$X_{1-p} \, Y_{1+p} \, Z_2 \, O_6$$

Z = tetrahedral site: Si, Al
Y = octahedral site: Mg, Fe^{2+}, Mn, Fe^{3+}, Al, Ti, (Ni, Cr)
X = 8-fold coordinated site: Ca, Na, (K), Li, Mg, Fe^{2+}, Mn

Pyroxenes of the enstatite–ferrosilite series are orthorhombic (orthopyroxenes or Opx); all others are monoclinic (clinopyroxenes or Cpx).

Pyroxenes are classified according to the filling of the X site X:

• ferromagnesian pyroxenes
• calcic pyroxenes

- sodic pyroxenes
- lithium pyroxenes

Ferromagnesian and calcic pyroxene (which are the most common), are represented in the trapeze $MgSiO_3 - FeSiO_3 - CaMgSi_2O_6 - CaFeSi_2O_6$ (Figure 3.22):

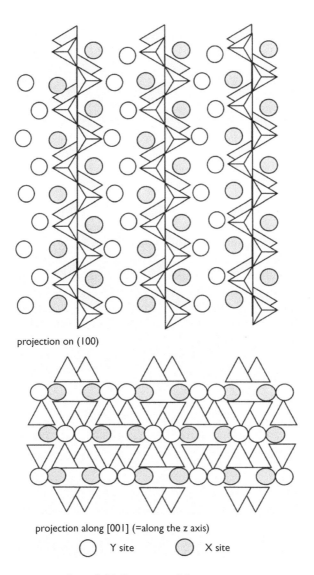

projection on (100)

projection along [001] (=along the z axis)

○ Y site   ◉ X site

Figure 3.21 Structure of the pyroxenes.

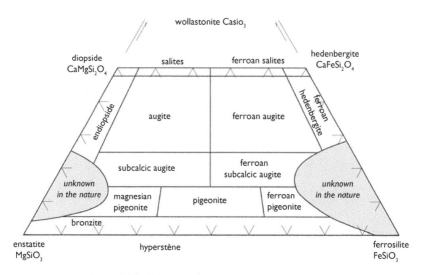

Figure 3.22 Calcic and ferro-magnesian pyroxenes.

Ferro-magnesian pyroxenes

Orthopyroxenes form the series: $MgSiO_3$ enstatite–hypersthene–ferrosilite $FeSiO_3$ (this last end member does not exist in nature). Protoenstatite is a polymorph of high temperature (above 1000°C) that is stable at relatively low pressure (less than 10kb). There are some very rare monoclinic polymorphs: clinoenstatite (volcanic rocks and meteorites) and clinoferrosilite (volcanic rocks). Bronzite is intermediate between enstatite and hyperstene but with a negative optic sign (hyperstene is optically positive); the name bronzite is nowadays considered a synonym of hypersten.

Pigeonite is a ferromagnesian calcic pyroxene, poor in calcium (5–15% of $CaSiO_3$ constituent). There is a miscibility gap with augite. Pigeonite is monoclinic. It is distinguished under the microscope from other clinopyroxenes by the angle of the optic axes 2V which is small (0–25–30°) (2V ranges from 25–62° in augite). Pigeonite is commonly transformed in igneous rocks by a fall of temperature into hypersthene with exsolution lamellae of augite ("inverted pigeonite").

Calcic pyroxenes

Diopside $CaMgSi_2O_6$–salites–***hedenbergite*** $CaFeSi_2O_6$–johannsenite $CaMnSi_2O_6$.series.

Augites $(Ca, Mg, Fe^{2+}, Al)Si_2O_6$ are very common minerals in basic igneous rocks (gabbros, dolerites, basalts) and in ultramafic rocks; they are somewhat less common in rocks of intermediate composition.

Introduction of aluminum in the network occurs by the substitutions:

$$Mg\ Si \Leftrightarrow Al^{VI}\ Al^{IV}$$
$$Mg\ Si \Leftrightarrow Fe^{3+}\ Al^{IV}$$

The corresponding end members are respectively $CaAl_2SiO_6$ (Tschermak molecule) and $Ca\ Fe^{3+}AlSiO_6$ (esseneite).

Al_2O_3 content of common augites of igneous rocks ranges from 1.5–4 wt%. High temperatures and high pressures favor the introduction of aluminum in the lattice.

Introduction of titanium in the lattice of pyroxenes occurs by substitution:

$$Mg\ Si_2 \Leftrightarrow Ti\ Al^{IV}_2$$

(corresponding end member $CaTiAl^{IV}\ O_6$).

In the above substitutions aluminum enters mostly in tetrahedral site.

Augites containing 1–2 wt% titanium are called titaniferous augites; those with a content of 3–5 wt% (or more) are called titanaugites.

Common augite contain 0.3–0.7 wt% of sodium in the Xsite. This content may reach up to 2–3% in the sodic augite.

Aluminous diopsides (also called *fassaites*,) calcic aluminous non sodic pyroxenes, are derived from diopside by the above substitutions. These minerals are distinguished from aluminous augites by their high calcium content (CaO around 25 wt%) so that in the structural formula Ca = 1 (in augite CaO content is generally below 21 wt% and Ca in the structural formula is less than 0.9).

The general formula of aluminous diopside is thus:

$$Ca\ (Mg, Fe^{2+})_{1-x}\ (Al^{VI}, Fe^{3+})_x\ Si_{2-x},\ Al^{IV}_x O_6$$

Some aluminous diopsides contain several percent of titanium.

Sodic pyroxenes

These form two groups that are distinguished by their chemical composition and occurrences:

1 sodic aluminous pyroxene whose formula is derived from that of diopside by substitution:

$$Ca\ Mg \Leftrightarrow Na\ Al^{VI}$$

Aluminum is then in the octahedral site.

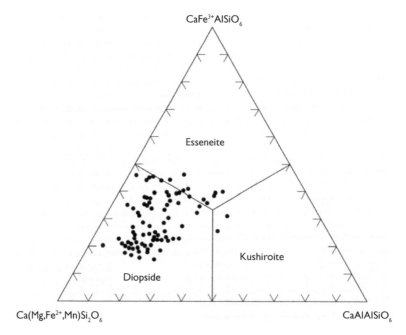

Figure 3.23 Fassaites: exemple of pyroxenes from skarns in Roumanie (after M-L Pascal, 2005, modified).

Jadeite is the $NaAlSi_2O_6$ end member; intermediate terms with diopside are called *omphacite*. Omphacites rich in ferric iron (intermediate between omphacite and aegirine) are called chloromelanites. These are minerals of high pressure metamorphism.

2   The iron-bearing sodic pyroxenes derived from diopside by substitution:

Ca Mg ⇔ Na Fe³⁺

The $NaFeSi_2O_6$ end member is *aegirine* (or acmite). The intermediate terms are called *aegirine augite*. Notice the fact that there is no solid solution between augite and aegirine and that the solid solution occurs between members of the diopside – hedenbergite series and aegirine. These are minerals of differentiated alkaline igneous rocks: volcanic (phonolites) and plutonic (nepheline syenites).

Spodumene

Lithium pyroxene is *spodumene* $LiAlSi_2O_6$. This mineral occurs in lithium granitic pegmatites with quartz, albite, lepidolite, beryl and petalite ($LiAlSi_4O_{10}$).

3.2.4.2 Stability of pyroxenes

Ferro-magnesian and calcic pyroxenes are minerals of high temperature, the stability of which is limited towards the lower temperatures by hydrated minerals, amphibole, talc, mica, etc. and/or carbonates. For instance:

$$5\ MgSiO_3 + H_2O \Leftrightarrow Mg_2SiO_4 + Mg_3Si_4O_{10}(OH)_2$$
enstatite + fluid $\Leftrightarrow$ forsterite + talc

(equilibrium temperature of about 650–700°C for pressures $P = P_{fluide}$ 1–3 kb; Kitahara et al. 1966).

$$5\ CaMgSi_2O_6 + H_2O + 3\ CO_2 \Leftrightarrow Ca_2\ Mg_5Si_8O_{22}(OH)_2 + 3\ CaCO_3 + 2\ SiO_2$$
diopside + fluid $\Leftrightarrow$ tremolite + calcite + quartz

(equilibrium temperature of about 450–550°C for P_{fluid} 1 kb, this temperature depends on the H_2O/CO_2 ratio; Metz, 1970, Skippen, 1974, Slaughter et al., 1975)

Pyroxenes can be stable at lower temperatures if the pressure of fluid (H_2O) is low.

Ferrosilite $FeSiO_3$ is stable only at pressures higher than 14.5kb. At lower pressure it is transformed into fayalite (Lindsley et al., 1964):

$$FeSiO_3 \Rightarrow Fe_2SiO_4 + SiO_2$$
ferrosilite $\Rightarrow$ fayalite + quartz

The diopside–hedenbergite system is relatively simple: it is a continuous solid solution whose phase diagram is similar for example to the forsterite–fayalite diagram of the olivines. It is somewhat more complicated towards the hedenbergite end member due to the fact that herdenbergite is replaced by iron-wollastonite at high temperature.

The $MgSiO_3$–$FeSiO_3$–$CaMgSi_2O_6$–$CaFeSi_2O_6$ system (pyroxene trapeze) is characterized by a miscibility gap between diopside–hedenbergite (and augite) series and the enstatite–ferrosilite (and pigeonite) series. Towards the magnesian end members, the solidus intersects the solvus that defines this gap.

This diagram is established for pure magnesian terms. The diagrams established for other ferro-magnesian and calcic pyroxenes show a similar topology. Introduction of iron in the system considerably lowers the temperatures. However, the phase diagram is poorly known for iron-rich terms.

At lower pressure, there is a eutectic between diopside and pigeonite; temperatures are also significantly lowered.

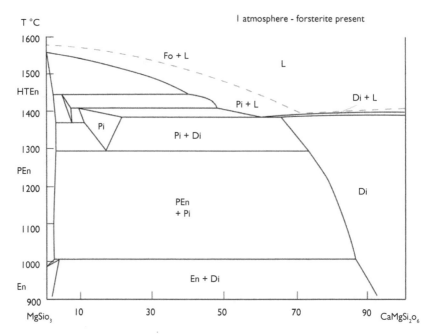

Figure 3.24 Diopside–enstatite system (after Carlson, 1988, modified).

L = liquid; Fo = forsterite; PEn = protoenstatite.
HTEN = phase analogous to high temperature enstatite.
En = enstatite; Pi = pigeonite; Di = diopside (solid solution = endiopside/augite).

Igneous petrology is largely based on the phase diagrams involving pyroxene–quartz–olivine–feldspars, to which we refer the reader.

Stability of jadeite is limited towards low pressure – and high temperatures – by the reaction (Figure 3.9):

$$NaAlSi_3O_8 \Leftrightarrow NaAlSi_2O_4 + SiO_2$$
Albite ⇔ jadeite + quartz

3.2.4.3   Occurrences of the pyroxenes

Igneous rocks

Pyroxenes are typically magmatic minerals that have crystallized at high temperature with low water pressure. Therefore these are mostly minerals of volcanic and hypabyssal rocks.

In plutonic rocks, the pyroxenes are abundant in the initial terms (gabbro, norite). They are rarer in differentiated terms (diorites) where they are replaced by amphibole. They are exceptional in granitoids. Hypersthene granitoids (charnockites) have crystallized in the granulite facies.

Pyroxenes of the *alkaline series* that occur in basic to intermediate rocks, augites close to the diopside–hedenbergite series or salites. These augite are richer in aluminium and titanium than in the subalkaline series, all the more aluminous and titaniferous as the series is more alkaline. In differentiated rocks, the pyroxene is aegirine and/or aegirine augite. In plutonic rocks (syenite, nepheline syenite, alkali granite), aegirine is commonly associated with riebeckite.

Hedenbergite appears with fayalite in rare alkali granites and quartz syenite, oversaturated in silica and highly under-saturated in alumina.

Basic and intermediate rocks of the *subalkaline series* contain two pyroxenes: an augite and calcium-poor pyroxene.

In the differentiated *lavas* (and hypabyssal rocks) of the *tholeiitic series*, this pyroxene is a pigeonite, forming both microlites and phenocrysts (in smaller quantities and more rare). In little or no differentiated lavas of these

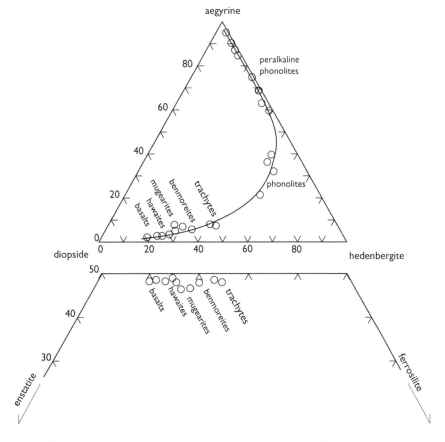

Figure 3.25 Pyroxenes of the Velay volcanic series, French Massif Central (unpublished data by E. Berger).

series, pigeonite is replaced by the enstatite/bronzite, because magnesian pigeonite crystallizes at higher temperatures (above 1300°C) than such magmas (which are about 1100 to 1200°C).

In *tholeiitic plutonic complexes* (anorthosite massifs, stratiform mafic–ultramafic complex like the Bushveld and the Skaergaard (Figure 3.26), pigeonite and augite of primary crystallization, formed at high temperature, have undergone changes during subsolvus progressive cooling producing exsolutions lamellae of orthopyroxene in augite and transforming the primary pigeonite into inverted pigeonite (hypersthene with exsolutions lamellae of augite). These stratiform complexes commonly contain accumulative pyroxenites (cumulates), like bronzitites and diopsidite.

The two pyroxenes of the *calcalkaline series*, volcanic or plutonic complexes (Sudbury, Giant Mascott) (Figure 3.26) are augite and hypersthene. In the plutonic complexes, augite, and to a lesser extent, hypersthene show the same type of exsolution as in tholeiitic plutonic complexes.

The diopside–enstatite phase diagram (Figure 3.24) shows that the tholeiitic series (with pigeonite) crystallizes at higher temperatures and lower water fugacities than the calc-alkaline series (with hypersthene).

Pyroxenes are, with olivine, the major constituents of ultramafic rocks: lherzolites (olivine + Opx + Cpx), harzburgite (olivine + Opx), wherlite (olivine + Cpx), websterite (Cpx + Opx) and pyroxenites.

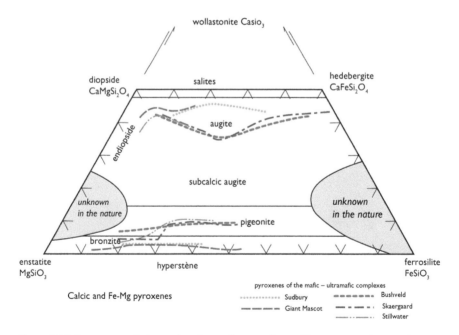

Figure 3.26 Pyroxenes of the mafic–ultramafic complexes (data compiled by M. Besson and unpublished data by the same author).

Occurrences of ultramafic rocks are:

- bodies of cumulative origin associated with basic–ultrabasic complexes, in particular, as more or less stratiform levels in Bushveld type layered complexes;
- usually small ultramafic intrusions, as the Pyrenean lherzolites;
- inclusions in alkali basalts (basanite, nephelinite) and kimberlites; it may be either cumulates, or preserved fragments of the mantle, or relicts of a partial melting of the mantle;
- large bodies, made mostly of harzburgites (and minor lherzolites), in Alpine type ophiolites.

The upper mantle is composed of lherzolite: olivine, orthopyroxene, clinopyroxene (diopside–endiopside) and a minor aluminous phase, spinel or, at higher pressure/depth, garnet.

In summary, most ultramafic rocks are:

- either cumulates formed from basic magmas;
- or fragments of the mantle emplaced by tectonics or inclusions in volcanic lava or kimberlites.

Stony meteorites (stones), chondrites and achondrites, which represent 90% of the meteorites, are composed of olivine, pyroxene and ore (less than 10%, as iron-nickel alloys and sulfides):

- iron-poor chondrites (L chondrites) with olivine and hypersthene;
- iron-rich chondrites (H chondrites) with olivine and bronzite;
- calcium poor achondrites with pigeonite or hypersthene;
- calcium-rich achondrites with diopside or augite.

Metamorphic rocks

Granulite facies (and that of pyroxene hornfels) are defined by the paragenesis orthopyroxene + calcic plagioclase. This paragenesis appears in the basic rocks. In the orthogneisses with a composition of granitoids, orthopyroxene associated with potassium feldspar is formed in the place of the biotite of common granites. Orthopyroxene is rare in aluminous paragneisses: association garnet + K-feldspar replace aluminium-rich biotite (see chapter 3.2.1d on micas). Clinopyroxene may be present in basic rocks of the granulite and pyroxene hornfels facies. It is rare in basic rocks of the amphibolite facies, where it is commonly replaced by hornblende.

Aluminous diopside also occurs in calcic-magnesian, aluminous, silica-undersaturated (Tschermak molecule is under-saturated in silica compared to anorthite) rocks of high temperature (pyroxene hornfels facies, granulite facies and high temperature amphibolite facies).

The *banded iron formations* that have undergone high grade metamorphism, contain iron-rich pyroxenes: orthopyroxene close to ferrosilite, hedenbergite, ferriaugite.

Clinopyroxenes of the diopside–hedenbergite series occur in *carbonate rocks*, impure marbles and calc-silicate-gneisses in a wide temperature range: the diopside isograd in these rocks corresponds roughly with that of aluminous silicates (garnet, cordierite, etc.) in metapelites.

Jadeite + quartz association defines a high-pressure sub-facies in *blueschist facies* (pressure over 7–8 kb). But in the absence of quartz, jadeite can be stable at lower pressures. Some rocks deficient in silica and alumina, in particular iron-rich formations in blueschist facies, may contain aegirine associated with an amphibole of the glaucophane–riebeckite series (for instance, Saint-Veran, Hautes Alpes, France).

Pyroxenes of the *eclogites* are omphacite. But the composition of those omphacites depends on the type of eclogite: the relatively low temperature eclogites in the areas of Alpine metamorphism are rich in jadeite component. Increase of temperature promotes the introduction of the Tschermak molecule. Eclogites in enclaves in alkali basalts contain aluminous diopside.

Metasomatic rocks

Pyroxenes are major components of the skarns. In skarns of high temperature (often at the magmatic stage) pyroxenes are commonly aluminous diopside. In most common hydrothermal skarns, pyroxenes belong to the diopside–hedenbergite–johannsenite series.

Aegirine appears in fenites, metasomatic halos around alkaline complexes, particularly carbonatites.

3.2.5   Olivine group

3.2.5.1   *Structure and chemical composition*

Minerals of the olivine group are nesosilicates which crystallize in the orthorhombic system. Isolated $(SiO_4)^{4-}$ tetrahedra are linked by divalent cations in octahedral position, occupying two non-equivalent sites. Hence the general formula:

X_2SiO_4
$X = Mg, Fe^{2+}, Mn, Ca, Ni, Cr, Ti$

Aluminium does not enter the tetrahedral.

The most important minerals in this group are *olivines*, which form a continuous series between the magnesian end member, Mg_2SiO_4, *forsterite* and the iron end member, Fe_2SiO_4, *fayalite*. The composition of olivine is noted by its proportion of forsterite, Fo%. Forsterite can contain up to

0.50 wt% nickel – it is the origin of New Caledonian type lateritic nickel deposits. Chromium levels are generally below 0.1 wt% except in olivines from komatiites where they reach 0.33 wt%. The manganese contents are lower than 0.50 wt% (except in certain alkaline acidic rocks where they reach 2 wt%). Peridot is a gem-quality forsterite.

Forsterite is the major constituent of the upper mantle. At greater depths, it is replaced by two polymorphs: beyond 410 km by wadsleyite (a sorosilicate) and beyond 520 km by ringwoodite (with spinel structure).

Magnesian olivines are an important constituent of mafic and ultramafic rocks and metamorphic impure dolomitic limestones. The iron-bearing olivine is rarer and occurs in differentiated igneous rocks of the alkaline and tholeiitic series.

Minerals of the fayalite–*knebellite* $FeMnSiO_4$–*tephroite* Mn_2SiO_4 series are rare and occurs almost only in the metamorphosed (iron) – manganese deposits.

Monticellite $CaMgSiO_4$ is a mineral of metamorphic limestones, skarns and some rare alkaline rocks, carbonatites, alnoites (melilite lamprophyre). It will be treated with calcic and magnesian minerals.

3.2.5.2   Stability of olivines

Melting point of forsterite is at 1890°C, that of fayalite at 1205°C. The phase diagram is that of the simple, continuous series type: early olivines are rich in magnesium and they are progressively enriched in iron during fractional crystallization.

The Mg_2SiO_4–SiO_2 system shows a peritectic point at 1560°C with reaction between olivine and the liquid to form enstatite (under a pressure of 1 atmosphere). At higher pressure there is a eutectic point between forsterite and (proto)-enstatite. The system Fe_2SiO_4–SiO_2 shows a eutectic point at 1177°C. The boundary between the two types of phase diagrams is poorly known.

Forsterite is therefore incompatible with quartz, whereas fayalite is.

Towards lower temperatures, stability of olivine is limited by the appearance of hydrated phases: ferro-magnesian amphiboles $(Mg, Fe)_7Si_8O_{22}(OH)_2$, serpentine $Mg_3Si_2O_5(OH)_4$, talc $Mg_3Si_4O_{10}(OH)_2$, brucite $Mg(OH)_2$ and/or carbonate (dolomite, magnesite). Formation of these phases is controlled not only by temperature but by the fugacities of H_2O and CO_2 and chemical potential of SiO_2.

3.2.5.3   Occurrences of olivines

Meteorites

Olivine is a major constituent of chondrites. Its composition varies depending on the type of chondrite: Fo 100–95 in enstatite E chondrites, Fo 84–80 in iron-rich H chondrites (bronzite), Fo 79–74 iron-poor L chondrites (hypersthene). The olivine compositions of achondrites are much more variable: Fo 96–66 for the ones with hypersthene, Fo 42–35 for the ones with augite. Pallasites are lithosiderites (stony-iron) with olivine (Fo 90–88) in an iron-nickel matrix.

Igneous rocks

It is estimated that the *mantle* olivines have compositions in the range of Fo 92. Lower forsterite contents suggest partial melting–differentiation phenomena.

Olivines from peridotites of *ophiolitic complexes* and alpine-type lherzolite intrusions have very constant compositions Fo 93–89 (mostly in the range Fo 92–91). Some rocks, in minor amounts, associated with these peridotites (some cumulative dunites, olivine pyroxenite indicating partial melting) show compositions ranging from Fo 88 to Fo 78.

Olivines in peridotite *inclusions in basalts and kimberlites* have compositions similar to those of alpine peridotites: Fo 90–91. In some cases, compositions in the range of Fo 88–83 have been found. In some enclaves of kimberlites compositions in the range of Fo 93 or more have been found.

Cortlandites are amphibole–olivine rocks, that are generally associated with mafic rocks (gabbros, diorites) in complex granitic batholiths.

Komatiites are ultramafic volcanic rocks formed from very magnesian magma: major olivine (Fo 90 or more), with chromian pyroxene, much subordinated anorthite and chromian spinel. The lava flows includes a thick cumulative basis (forsterite-rich olivine, chromian spinel in primary magma) and, with sharp boundaries, a thinner upper part in which the rapid crystallization causes a characteristic dendritic texture (spinifex).

Olivines are major minerals of *volcanic and hypabyssal rocks*, mostly in basic rocks and less frequently in the intermediate and acidic rocks.

Generally the compositions of olivine are consistent with the hypothesis of fractional crystallization with an evolution from forsterite-rich terms to terms richer in fayalite. This trend is observed in the same rock by the zonation of the phenocrysts and differences between the phenocrysts and microlites, or between the various rocks of the same magmatic suite. However in some rocks, such as olivine crystals, are xenocrysts that do not follow from these laws of fractional crystallization.

The picrites and oceanites are cumulative olivine basaltic; they may belong either to alkaline or tholeiitic series.

In the *subalkaline series*, there is *a priori* only a single generation of olivine: early phenocrysts, more or less zoned, which tend to be resorbed, corroded by the magma and/or react with it, to eventually form hypersthene.

In the tholeiitic series (and the MORB), olivine is common in poorly differentiated terms (olivine tholeiites, basaltic picrites, basalts): its composition varies from Fo 90/89 to Fo 82/73. It is missing in the intermediate terms. Iron-rich olivines only appear in very differentiated terms, if any.

In *mafic-ultramafic stratiform plutonic complexes* of Bushveld/Skaergaard type, olivine appears in the poorly differentiated and cumulative basal parts of the complexes: Fo 90–86 in bronzitites, harzburgites, dunites of the basal zone of the Bushveld complex; Fo 67–53 in the gabbros of the lower part of the Skaergaard intrusion. It is lacking in the middle parts of these complexes. Iron-rich olivines appear in the upper parts: Fo 45–5 in the ferrogabbros, ferrodiorites, hortonolites (=iron-rich olivine dunite) from the upper zone of the Bushveld complex; Fo 40–0 ferrogabbros and ferrodiorites from the upper zone and border group of Skaergaard.

In the calc-alkaline series, olivine only appears in very little evolved terms (basalt, basaltic andesite).

In *alkaline volcanic rocks*, there are two generations of olivine: early phenocrysts and microlites.

The most common alkaline series are of basalt/basanite–hawaiite–mugearites–benmoreites–trachyte–phonolite association type. Most times olivine is important in basalts/basanites and hawaiites and ceases to crystallize in more evolved terms. Its composition varies in these rocks from Fo 85–80 to about Fo 55–50. In some series, olivine is also present in the most evolved terms, trachytes and even up to phonolites. Compositions are up to Fo 35–27. Fayalite (Fo 15–5) is present in some alkaline rhyolites (pantellerites, comendite). Fayalite (For 17–0) also appears in *differentiated alkaline plutonic rocks*, silica-saturated or not (nepheline syenite, quartz syenite, alkali granite); it is associated with sodic amphiboles, rich in iron (riebeckite–arfvedsonite) or not so rich (hastingsite) and with hedenbergite.

Magnesian olivines are also important in the *peralkaline basic rocks of the nephelinite family* (nepheline basalts, leucitites, melilite basalts) (Fo 92 to 70 in differentiated terms) and in the kimberlites (For 95–85) and lamproites (ultrapotassic hypabyssal and extrusive rocks with forsterite, leucite, sanidine, phlogopite).

Olivine is a common constituent of some rocks of *alkaline mafic-ultramafic complex*. Compositions range from Fo 93–77 in carbonatites, alnoites (melilite lamprophyres) and melilitites.

Metamorphic rocks

Forsterite appears in high grade regional or contact metamorphism of impure dolomitic limestones, dolostones and in skarns.

Fayalite is common in metamorphosed iron formations, especially in the banded iron formations (BIF).

Prograde metamorphism of serpentine produces olivine (with talc, chlorite, tremolite, magnesian amphibole). Hydration of pyroxenites (Cpx + Opx) produces the paragenesis olivine + calcic amphibole.

3.2.5.4   Alterations of olivine

Iddingsite is an alteration product of olivine of volcanic rocks; it appears as a reddish (orange, red brownish, brown, black) fringe located at the periphery of the crystals or along their fractures. Iddingsite is an aggregate made of cryptocrystalline goethite, hematite, smectite (saponite, montmorillonite) and sometimes quartz, calcite and magnetite. It is more or less pleochroic depending on the more or less ordered orientation of goethite and the sheet minerals. The transformation of olivine into iddingsite involves oxidation, hydration and some leaching of magnesium. It is most often a deuteric alteration due to the effect of fumaroles on the lava.

Olivine is also altered into greenish cryptocrystalline aggregates (called bowlingite), made of chlorite, smectite (saponite), and possibly serpentine, talc, brucite, mica. etc. This may be a deuteric alteration, like iddingsitization, but also a supergene alteration.

Chlorite, more or less largely crystallized, often with Prussian blue interference colors, may replace olivine.

Serpentinization is a very common and spectacular alteration of ultramafic rocks.

3.2.6   Serpentine and serpentinization

Serpentines are phyllosilicates composed of kaolinite-type TO layers with a tetrahedral layer and octahedral layer. The formula of serpentine is:

$$Mg_3Si_2O_5(OH)_4$$

with very low amounts of iron replacing magnesium and, traces of Cr, Mn, Co, Ni.

The most common variety of serpentine is *antigorite*. The bastite is an antigorite that appears as a more or less complete pseudomorph of orthopyroxene. *Chrysotile* is the fibrous variety, used once to asbestos. Lizardite is a very fine grained, scaly variety. Some varieties have qualities of semiprecious stones imitating jade ("bowenite", "serpophite").

Experimental data indicate temperatures of transformation of forsterite into antigorite (with talc or brucite) in the range from 450–600°C, depending on water pressure. But the temperatures of serpentinization in natural systems are apparently lower: antigorite is formed at temperatures of about 220–480°C, lizardite and chrysotile at temperatures of about 85–115–250°C.

Serpentinization is the transformation of ultramafic rocks, with olivine and orthopyroxene (dunite, harzburgite, lherzolite) in rocks formed of serpentine, magnesian chlorite, talc, brucite, magnetite, magnesite and/or dolomite. This transformation involves hydration and is likely accompanied

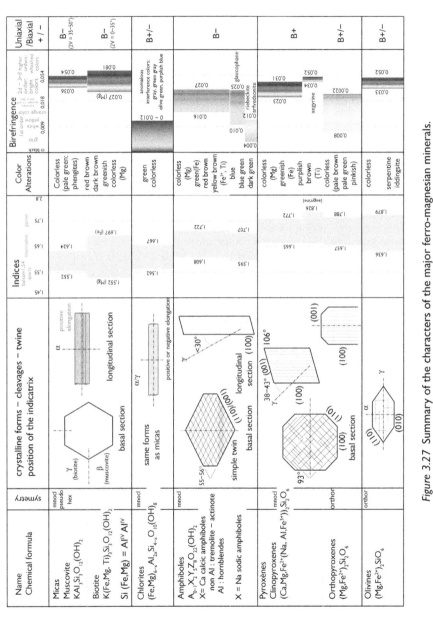

Figure 3.27 Summary of the characters of the major ferro-magnesian minerals.

by an increase in volume. Calcium may come from the alteration of clinopyroxene – and/or possibly from a carbonatization.

Ultramafic rocks of ophiolite complexes are generally more or less completely serpentinized. Serpentinization is attributed to hydrothermal metamorphism of the rocks of the oceanic crust and of the associated rocks of the upper mantle. Metamorphism of greenschist or blueschist facies may be superimposed on the hydrothermal metamorphism contemporaneous with the formation of these rocks. Ultramafic rocks of greenstone belts of Archean and Proterozoic age are also largely serpentinized. Serpentinization is probably less frequent and intense in the layered mafic–ultramafic complexes.

3.3   ALUMINOUS MINERALS

These minerals are mainly metamorphic minerals of metapelites (and paleo-alterites) initially more or less rich in clay (and not calcareous). They also occur in peraluminous igneous rocks: granodiorites, granites, aplites, pegmatites, etc. and in some metasomatic rocks whose peraluminous character is due to the leaching of alkalis and calcium (greisens, "secondary hydrothermal quartzites", etc.).

Other peraluminous minerals, muscovite, biotite and chlorite, have been previously presented. Spinel $((Fe, Mg)Al_2O_4)$ and tourmaline are dealt with accessory minerals. Aluminous minerals are treated here:

- alumina silicates: andalusite, sillimanite, kyanite;
- most commonly ferriferous aluminous silicates: garnet (almandine), staurolite, chloritoid;
- mostly magnesian aluminous silicates: cordierite magnesiocarpholite, sapphirine;
- topaz, corundum and beryl, minerals whose mode of occurrence is somewhat different and that could be classified with accessory minerals;
- pyrophyllite, diaspore and aluminum hydroxides, gibbsite and boehmite; the latter two are minerals of sedimentary rocks and are not determined under the petrographic microscope, and therefore not treated here.

3.3.1   Structure and chemical composition

3.3.1.1   Alumina silicates: andalusite, sillimanite, kyanite

There are three natural polymorphs of formula Al_2SiO_5: sillimanite polymorph of high temperature, kyanite of high pressure and relatively low temperature and andalusite of low temperature and low pressure. The composition of natural minerals does not differ very much from the ideal formula. The only notable substitutions are those of aluminum by ferric iron in andalusite (up to 2 wt% of Fe_2O_3), and to a lesser extent in the kyanite, and manganese in andalusite. The iron-rich andalusites have a pink color under the microscope, those rich in manganese and iron (viridine) a green color.

Alumina silicates consist of chains of octahedra joined by an edge, containing aluminum. These chains are linked by isolated silica tetrahedra (these minerals are nesosilicates) alternating with sites containing the other half of aluminum the coordination of which varies in these various polymorphs:

- the pressure tends to favor the phases where aluminum has a 6-fold coordination. The structural formula of kyanite is $Al^{VI}Al^{VI}SiO_5$ the symmetry of this mineral is triclinic but the network is close to a face-centered cubic lattice;
- high temperatures favor the phases where aluminum has a 4-fold coordination: sillimanite, high-temperature polymorph of orthorhombic symmetry, has the formula $Al^{IV}Al^{IV}SiO_5$;
- half of the aluminum of andalusite have a 5-fold coordination: $Al^{VI}A^{IV}SiO_5$; this mineral has orthorhombic, almost tetragonal, symmetry.
- Only the curve of equilibrium between sillimanite and kyanite is precisely known. The other curves andalusite–sillimanite and andalusite-kyanite, are poorly known so that different values for the position of the triple point have been proposed: 350 to 600°C, 3–8 kb. Recent data suggest 500 ± 50°C and 4 ± 0.5 kb (Figure 1–9).

3.3.1.2 Aluminous garnets

Garnets are made of a three-dimensional network of tetrahedra (site Z) and octahedra (Y site). Octahedra and tetrahedra alternate and share an oxygen to each apex. This structure makes room for distorted cube-shaped sites (X sites) of 8-fold coordination. Thus the formula of garnet is:

$X_3 Y_2 Z_3 O_{12}$
$X = Fe^{2+}$, Mg, Mn Ca, (Y)
$Y = Al, Fe^{3+}, Cr, (V, Mn^{3+}, Ti, Zr, Fe^{2+})$
$Z = Si$

In most garnet silicon is the only element entering in tetrahedral site. In titanium or zirconium garnets aluminum can enter this site. Tin enters in this site in some rare calcium garnets.

Garnets are traditionally divided into aluminous garnets and calcic garnets. These two groups differ not only by their chemical composition but also by their occurrences.

The end members of aluminous garnets (pyralspites) are:

Pyrope $Mg_3 Al_2 Si_3 O_{12}$
Almandin $Fe_3 Al_2 Si_3 O_{12}$
Spessartine $Mn_3 Al_2 Si_3 O_{12}$
Grossular $Ca_3 Al_2 Si_3 O_{12}$

Although the grossular is also part of the group of calcium garnets and is usually treated with them, it is here associated to the group of pyralspites because they always contain a larger or less proportion grossular molecule (usually less than 10%, sometimes much more).

Aluminous garnets are common minerals in metamorphic rocks, particularly in rocks deriving from pelites, and in peraluminous igneous rocks. They do not show other substitutions than the ones between the elements in X site, mostly Fe^{2+}, Mg, Mn and Ca.

Garnets from kimberlites and some peridotites may contain a significant proportion of chromium, defining a series between the pyrope end member and a knorringite $Mg_3 Cr_2 Si_3 O_{12}$ end member.

3.3.1.3 Staurolite

Staurolite is a nesosilicate formed of Al_2SiO_5 slabs the structure of which is identical to those of kyanite, alternating with iron-layers of approximate composition $AlFe_2O_2(OH)_2$. This analogy with the structure of kyanite makes that these minerals occur quite often in epitaxic association. The aluminum sites are not fully occupied. Thus the precise formula of staurolite is not exactly known:

$$(R^{2+})_{<2} \ (R^{3+})_{about\ 9} \ Si_{about\ 4} \ O_{22} \ (OH)_{about\ 4}$$

where R^{2+} = Fe^{2+}, Mg, Zn, (Co, Li)
      R^{3+} = Al, (Fe^{3+}, Ti, Cr)

Staurolite contains mostly iron as R^{2+}, with moderate substitution of iron by magnesium. It contains very commonly zinc in small quantities. Substitutions of Al are very small. So the simplified formula of staurolite is:

$$(Fe, Mg, Zn)_2 \ Al_9Si_4O_{22}(OH)_2$$

Staurolite is typically a mineral of medium grade metamorphism: its stability domain ranges from about (500)–550°C to 650–700°C. At higher temperatures it is destabilized into garnet (almandine) + aluminum silicate + water. Towards lower temperatures, the reaction of formation of staurolite varies with initial parageneses (an example will be developed below). Its stability is limited to high pressure values of around 16–17 kb. Staurolite is rare at very low pressures. The paragenesis staurolite–cordierite–biotite–muscovite is known in many places (although some authors deny it for theoretical reasons).

3.3.1.4 Chloritoid

Chloritid is nesosilicates made of sheets parallel to the plane (001) (perfect cleavage plane) connected by isolated $(SiO_4)^{4-}$ tetrahedra. Each sheet consists of a compact assemblage of octahedra. There are alternating layers of Al_6O_{16} corundum-type layers and $M_4 Al_2O_4(OH)_8$ brucite-type layers. M can be Fe^{2+}, Mn, and Mg. In this brucite-type sheet, ferric iron can replace small amounts of Al, the introduction of ferric iron is accompanied by an OH deficiency. Hence the general formula of chloritoid is:

$$(Fe^{2+}, Mg, Mn)_2 (Al, Fe^{3+})Al_3O_2(SiO_4)_2(OH)_4$$

The chloritoid is a metamorphic mineral of low to medium temperature, stable at temperatures up to 600–650°C (first half of amphibolite facies).

Ottrelite, manganoan equivalent of chloritoid, appears in areas of low to very low levels metamorphic grade in manganese-rich levels (for instance in the Ardennes). The percentage Mn/(Fe + Mg + Mn) can reach 60%.

A (very) low pressure (and low and medium temperatures), only the iron chloritoid is stable. So this mineral appears only in the rocks of particular chemistry, both aluminum- and iron-rich and poor in alkali: for instance metamorphosed lateritic paleosols or bauxites (and the sediments derived from such paleoalterites).

At higher pressures the range of compositions of chloritoid widens: Mg/(Mg + Fe + Mn) ratio can reach 40% and even, exceptionally, 74%. Chloritoid appears in more varied rocks. It is a common mineral in metapelites of blueschist facies.

3.3.1.5 Cordierite

Classically, cordierite is considered as a cyclosilicate: $(Si_6O_{18})^{12-}$ rings in which part of the silicon atom is replaced by aluminum, connected by isolated $(SiO_4)^{4+}$ tetrahedra, where also a part of the silicon atoms is replaced by aluminum. This structure of rings linked by tetrahedra, makes that cordierite is sometimes considered a tectosilicate. The structure gives place to the octahedral sites occupied by Mg and Fe^{2+}. The common cordierites are mostly magnesian, with a limited Mg ⇔ Fe substitution.

There are two polymorphs of cordierite: the high temperature form, indialite, is hexagonal and the lower temperature form is orthorhombic, pseudohexagonal. The transition between the two forms is around 1450°C for the pure magnesian end member; the presence of iron lowers this temperature. Nevertheless this range of temperature is well above the temperatures of geological phenomena: the common cordierites are orthorhombic. In them, the Si/Al ratio hardly deviates from the value 5/4. The formula of cordierite is:

$$(Mg, Fe)_2 Al_4Si_5O_{18}$$

The stacking of $(Si, Al)_6O_{18}$ rings gives place to large channels that can provide lodging for water molecules or other fluids. This water is not connected to the lattice and should not be considered as a water of constitution, but as zeolitic water. Nevertheless, water is always present in analyses of cordierite. On the other hand, cordierite is most commonly altered into a yellowish isotropic product called pinite, that is mostly chlorite and/or muscovite. It is difficult to analyze cordierite with the microprobe due to this zeolitic water and such alterations.

The open structure of cordierite (tetrahedral sites, rings) means that this mineral is unstable at high pressure: at the higher pressures, cordierite is replaced by garnet + alumina silicate, or the association cordierite + K-feldspar is replaced by biotite (phlogopite) + alumina silicate. The upper limit of stability of cordierite is around 6–7kb. Some authors consider much lower pressures.

Cordierite is a mineral medium high temperature. Its melting point (1450°C for the magnesian end member) lies beyond the field of geological phenomena. At high temperatures, it may be replaced by associations of spinel + quartz or sapphirine + quartz.

3.3.1.6 Magnesiocarpholite

Magnesiocarpholite is a fibrous inosilicate, the structure of which is quite similar to that of orthopyroxene and whose formula is $(Mg, Fe)Al_2Si_2O_6(OH)_4$. The iron end member is ferrocarpholite. The manganoan end member of the series is carpholite, a rare mineral of pegmatites and metamorphic manganese-rich levels. Carpholite was described as early as 1817; but it was not until the use of the microprobe that magnesiocarpholite has been described (Goffé et al., 1973).

The range of temperature of stability of magnesiocarpholite seems limited (350–600°C?). It is to replaced at lower pressures by chlorite:

magnesiocarpholite + pyrophyllite $(Al_2Si_4O_{10}(OH)_2)$ + H_2O
⇔ chlorite $[(Mg, Fe)Al_4Si_3O_{10}(OH)_8)]$ + quartz

Magnesiocarpholite appears in metapelites in the high-pressure metamorphism with phengite–chloritoid–lawsonite–phyrophyllite. It seems characteristic of the lower blueschist facies (with lawsonite, without jadeite).

3.3.1.7 Sapphirine

Sapphirine is an inosilicate composed of complex chains of tetrahedra T_6O_{18} connected by bands of octahedra M_7O_2, an additional octahedral site connects these bands. The general formula of sapphirine is $M_7(M)O_2\,T_6O_{18}$, where site T is the occupied by Si or Al and M site by Mg, Al, Fe^{2+}, Fe^{3+}.

Sapphirines therefore derive from the formula $Mg_8Si_6O_{20}$ by substitutions Si Mg $\Leftrightarrow$ Al^{IV} Al^{VI}, and to a lesser extend Mg $\Leftrightarrow$ Fe^{2+} and Al^{VI} $\Leftrightarrow$ Fe^{3+}. Sapphirine is deficient in silica relatively to cordierite.

Sapphirine is a high temperature mineral (750–800°C) which is replaced at very high pressures (17–25 kb) by pyrope.

3.3.1.8 Topaz

$Al_2SiO_4(F, OH)_2$ is a mineral close to andalusite (especially in its optical properties). It is made of double chains of octahedra linked by isolated $(SiO_4)^{4-}$ tetrahedra. The octahedral sites are occupied by aluminum. Two of the six anions occupying the apexes of the octahedra are occupied by F or OH, the others by oxygen. The chemical composition of topaz is very uniform. The ratio F/F + OH seems always higher than 70%.

Experimental data suggest that topaz is stable within a temperature range of about 500–850°C.

3.3.1.9 Beryl

$Be_3\,Al_2Si_6O_{18}$ has a structure similar to that of cordierite: Si_6O_{18} rings connected by octahedra containing Al and distorted tetrahedra containing Be. As in cordierite, there are channels that can contain water and alkali (Na, Cs). The presence of trace elements either by substitution of Al or in the channels induces remarkable colorations (in macroscopic samples) of some beryls: Cr^{3+} and/or V^{3+} (emerald), Fe^{3+} (heliodor, golden yellow), Fe^{2+} (aquamarine, light blue green), Mn^{2+} (morganite, pink). Under the microscope, beryl is colorless.

3.3.1.10 Corundum

Al_2O_3 is made of layers of oxygen in an approximately hexagonal close packing, that provide intermediate sites occupied by aluminum; only 2/3 of these sites are occupied. The red color of ruby is due to the presence of chromium in substitution to aluminum, the blue color of sapphires of that of iron and titanium. Minute inclusions of rutile oriented at 60° (sagenite twin) produce the effect of asterism in star sapphires.

Corundum is under-saturated in silica and alkalis. Its high melting point (2000–2050°C) denies that it can be of magmatic origin.

3.3.1.11 Pyrophyllite

$Al_2Si_4O_{10}(OH)_2$ is a dioctahedral phyllosilicate formed by sheets similar to those of muscovite (two tetrahedral layers (Si) surrounding an octahedral layer (Al)). The sheets are linked by hydrogen bond; this very weak bond explains the disordered stacking of the sheets, the good cleavage and the low hardness of this mineral. Iron and manganese can be substituted in very small amounts of aluminum.

This is a mineral of low temperature. Its stability is limited towards the low temperatures (around 300°C) by the reaction:

kaolinite + quartz = > pyrophyllite + H_2O

and toward the higher temperatures (350–420°C) by

pyrophylite = > quartz + alumina silicate + H_2O
pyrophyllite + chlorite = > chloritoid + quartz + H_2O
pyrophyllite + calcite = > margarite + quartz + H_2O + CO_2

3.3.1.12 Diaspore (gibbsite and boehmite)

Diaspore α-AlO(OH) (with very low quantities of iron and manganese in substitution to aluminum) is formed of layers of oxygen in a hexagonal close packing; aluminum occupies 8-fold coordinated sites between the layers, forming strips of octaedra. **Boehmite** β-AlO(OH) is a polymorph of diaspore: double sheet of octaedra with aluminium at their centers where the oxygens are in a cubic packing relationship (instead of hexagonal close packing in diaspore). In **gibbsite** Al(OH)$_3$ the hydroxyls form sheets with hexagonal close packing.

Gibbsite and boehmite are mineral of supergene alteration. They are in particular, the major constituents of bauxites. Their very fine grain does not allow their determination under the microscope. They will not be further dealt here.

Diaspore appears in the alterites (soils, bauxites). It is formed as soon as diagenesis by recrystallization of gibbsite. It is a mineral of metamorphism of the same peraluminous rocks. It destabilizes corundum to high temperatures.

3.3.2 Occurrences

3.3.2.1 Metamorphic rocks

Contact metamorphism

Contact metamorphism around igneous rocks, and particularly around, granitoids, commonly forms spotted schists and hornflels with nodules of cordierite and andalusite (often with the chiastolite habit). It is less common that the metamorphism is strong enough so that sillimanite appears. Corundum is very rare; kyanite and staurolite are exceptional.

Regional metamorphism of metapelites

Muscovite, chlorite, biotite, andalusite, kyanite, sillimanite, garnet, staurolite, cordierite, chloritoid (and more rarely magnesiocarpholite) as well as plagioclase, are common metamorphic minerals of the metapelitic rocks. Isograds, defined by the appearance (or disappearance) of such minerals, and the paragenesis (mineral association in equilibrium) in which they are involved, are used to accurately characterize the regional metamorphism.

The appearance of these minerals is controlled by physical conditions (Figure 1.9). Chlorite, chloritoid and magnesiocarpholite are low temperature minerals; andalusite, kyanite, garnet, staurolite, cordierite, biotite, muscovite are characteristic of a medium-grade metamorphism; garnet, cordierite, sillimanite, sapphirine are stable at high or very high temperatures. Cordierite is a mineral of low pressure. Garnet is rare at low pressure, besides in the iron- and manganese-rich rocks. Staurolite is unlikely to appear at very low pressure. Magnesiocarpholite and kyanite are minerals of medium to high pressure. Chloritoid is more common in areas of medium to high pressure than in areas of low pressure.

The chemical composition of these minerals, and particularly the ration between iron and magnesium) allows the following classification:

- minerals without iron nor magnesium: alumina silicates, muscovite. Strictly speaking, muscovite should not be classified in this group, as it is generally more or less phengitic, but it can be considered as a first approximation a purely aluminous muscovite;
- mostly iron-bearing minerals: almandine garnet, staurolite, chloritoid;
- mostly magnesian minerals: cordierite, (magnesiocarpholite);
- (ferromagnesian) minerals with large variation of the iron–magnesium ratio biotite, chlorite, phengite.

At a given grade of metamorphism (that is at given physical conditions), the presence in a rock of any of such mineral, depends essentially on the proportions of iron, magnesium, aluminum and potassium.

Plagioclase is generally present; but, the balance between Fe, Mg, Al and K, only consider the aluminum that is not fixed in the plagioclase, which can be expressed as Al–Na–2Ca. Ilmenite is often present in metapelites and strictly speaking, this balance should take account of the iron fixed in ilmenite ($FeTiO_3$). It may be difficult to assess the proportion of iron contained in ilmenite: either one considers that all the titanium is contained in ilmenite and the parameter to be considered is Fe-Ti (but there is some titanium fixed in the biotite); another way is to consider that the amount of iron fixed in ilmenite is negligible and the parameter to be considered in this balance is the total Fe.

A low to medium grade of metamorphism, muscovite is almost always present in metapelites. It is therefore most times considered that muscovite is in excess: when studying the formation of minerals the balance between Al,

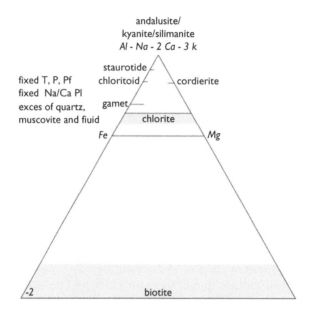

Figure 3.28 Thompson's diagram (1957).

Fe and Mg must take into account only the aluminum that is not contained in muscovite (Al-3 K) (nor in the plagioclase).

This reasoning leads so to reduce the discussion of metamorphic minerals in metapelites to a Fe/Mg/Al-Na-2Ca-3K diagram: the diagram of Thompson (1957) (Figure 3.28). This is a convenient way to represent the various possible parageneses at a given metamorphic grade and thus to define precisely metamorphisc facies.

The rigorous reasoning based on the phase rule indicates that the diagrams constructed to represent various metamorphisc facies, must satisfy the following conditions: fixed physical conditions (temperature, pressure, fluid pressure), fixed composition of the plagioclase (Na/Ca in plagioclase), excess of quartz, muscovite and fluid. Any change in one of these parameters will more or less change the diagram.

Note also that this diagram does not take account of manganese, an element that may be important in garnet.

A suite of (prograd) metamorphic facies may be so represented by a series of Thompson's diagrams. Comparison between these diagrams allows the discussion of the reactions occurring at the isograds.

There are many well documented examples that show the wide variety of facies and their succession. It is not sensible to theoretically define a standard succession of facies and isograds. Each case deserves a separate study.

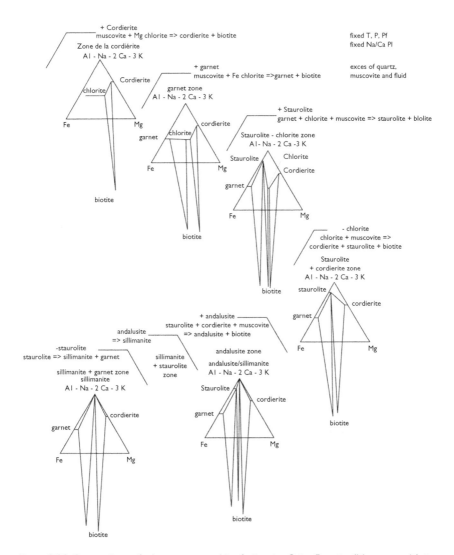

Figure 3.29 Succession of the metamorphic facies in Saint-Ponais (Montagne Noire, Hérault, France).

We give in Figure 3.29 the example of an intermediate low pressure metamorphism, observed in Saint-Ponais, on southern slope of Montagne Noire (France, Demange and Gattoni, 1976, Demange, 1982). Diagrams are built in actual size from mineral analyses with the microprobe.

The pattern of the Thompson's diagram established for a given zone of metamorphism, is only valid if the following conditions are fulfilled: fixed temperature, pressure, fluid pressure, fixed composition of the plagioclase. The phase rule indicates that any change in any of these parameters induces a different pattern. A significant change in temperature, pressure, etc. produces the appearance or disappearance of a new phase: this is what occurs at an isograd; and the topology of the corresponding diagram is changed. A moderate change of these parameters does not change the topology of the diagram but shifts the tie-lines that connect the associated minerals.

Consider for example the paragenesis biotite–chlorite–garnet–(quartz–muscovite), which is critical for the garnet zone in Saint-Ponais (Montagne Noire) (Figure 3.30a). Between the garnet isograd and the following staurolite isograd, as the temperature increases the compositions of these associated minerals varies and become progressively more magnesian with the increasing temperature: the diagram gradually changes by shifting the tie-lines.

Similarly if we consider the association biotite–chlorite–cordierite–(quartz–muscovite) in the cordierite zone (between the cordierite isograd and that of andalusite) in the Agly massive (Eastern Pyrenees) (Fonteilles, 1970), in the same way, with increasing temperature, the tie-lines shift: the associated minerals become progressively richer in iron (Figure 3.30b).

Note that, beyond the andalusite isograd, these trends of shifting of the tie-lines are inversed: biotites of the paragenesis biotite–garnet–staurolite or

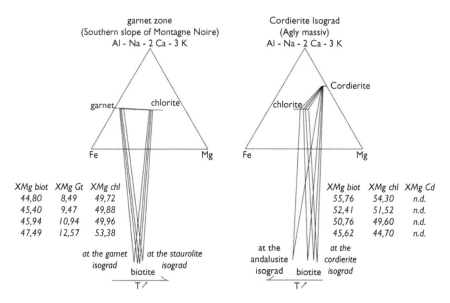

Figure 3.30 Shifting of the tie lines with temperature (Demange, 1982 and unpublished data)($XMg = 100\ Mg/Fe + Mg$).

andalusite/sillimanite become increasingly more ferroan when the temperature increases. Similarly, the biotite of the paragenesis of biotite–cordierite–sillimanite/andalusite become increasingly magnesian with increasing temperature.

In order to show the variations of the topology of Thompson's diagram with pressure, let us compare two rocks of the same chemical composition that were formed at equivalent temperatures. Both are from Montagne Noire, but one comes from the region of Murat-sur-Vèbre, a domain of low pressure metamorphism (with the isograds: + cordierite + andalusite + sillimanite, and where garnet or staurolite are practicallyabsents); the other one comes from the region of Labastide Rouairoux, a domain of medium pressure metamorphism (with the isograds: + garnet + staurolite + kyanite + sillimanite) (Figure 3.31).

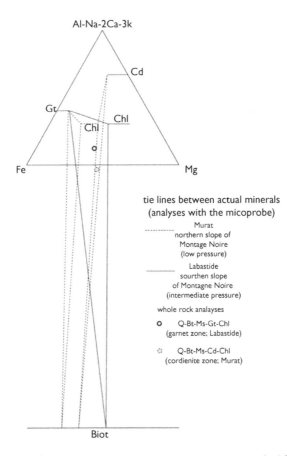

Figure 3.31 Shifting of the tie lines with pressure: two regions with different types of metamorphism, low pressure and intermediate pressure.

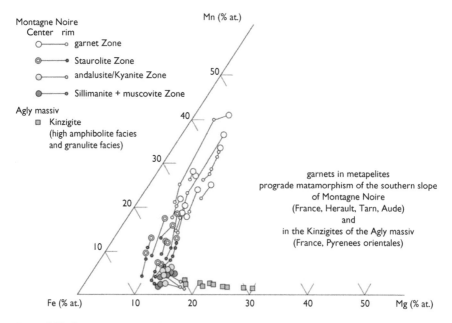

Figure 3.32 Garnets in prograde metamorphism: example from southern slope of Montagne Noire and granulite massif Agly (Demange, 1982 and unpublished data).

The diagram shows that the tie-lines of the garnet–biotite–chlorite paragenesis in the domain of medium pressure metamorphism, cross the tie-lines of the cordierite–biotite–chlorite paragenesis in the domain of low pressure metamorphism: metamorphism of rocks of the same chemical composition produces cordierite schists at low pressure and garnet schists at medium pressure.

Role of manganese in the appearance of garnet in metapelites

Thompson's diagram does not consider manganese but garnet often contains a significant amount of this element. It is often assumed that garnet is stabilized by the presence of manganese. This assertion must be nuanced and discussed by considering the composition of the "normal" garnets; such "normal" garnets, are the ones that the Thompson's diagram can explain without requiring the presence of manganese (at a give chemical composition of the rock and a type of metamorphism).

In a prograde metamorphism, "normal" garnets show the following evolution: the first garnets that appear at the isograd are rich in manganese (and calcium). With increasing metamorphism, they become more and more rich in iron, at the same time the ratio of magnesium/iron increases slightly. The fact that the manganese content of garnet decreases in prograde metamorphism, is due to the fact that the amount of garnet increases and that the manganese enter into other phases (biotite and probably ilmenite). At higher metamorphic grade (deep amphibolite facies and granulite facies), garnets are poor in manganese, and become more and more magnesian, with a very strong increase of the magnesium/iron ratio.

When garnet appears in an association that cannot be taken in account by Thompson's diagram, like, for instance, an association garnet–staurolite–andalusite–(biotite–muscovite) in the andalusite zone, it appears that the garnet does not fall within the previous trend, but is substantially richer in manganese than "normal" garnets. In this case (only), it may be accepted that these garnets are stabilized by manganese.

In the case of different types of metamorphism, the trends of the evolution of garnets with increasing matamorphism, are similar but shifted. With an increasing ratio P/T, this shift of the main trend is shifter both towards stronger Mg/Fe ratios, and lower contents of manganese.

Thompson's diagram indicates that the presence of garnet is mainly controlled by the Fe/Mg ratio of the host rock. Actually, at a given metamorphic grade and type, the presence of garnet is determined the Fe/Mg/Mn ratios. We can define for a given type of metamorphism, the fields of the rocks where garnet is present or not, as a function of this parameter. The diagram FeO/MgO *vs* FeO/MgO in whole rock (Figure 3.33) compares the boundaries of the field of rocks containing garnet and rocks without garnet in different metamorphic domains: low pressure metamorphism (Agly, Albères), intermediate low pressure metamorphism (Canigou massiv, southern Montagne Noire), high pressure metamorphism (Vanoise massiv) and very high pressure metamorphism (Gran Paradiso massiv). At low pressure (Agly massiv), garnet is restricted to rocks rich in both iron and manganese. A higher pressure garnet appears in rocks less rich in iron and/or manganese. This result is in accordance with the data on the shift of the tie-lines in the Thompson's diagram as a function of pressure (pictured in Figure 3.31).

Calcium in the garnets of non-calcareous metapelites

In non-calcareous pelitic rocks, calcium is essentially contained in the clastic plagioclase. Calcic plagioclase are not stable at low temperatures and, in the condition of a weak metamorphism, the garnet is the main bearer of calcium. The CaO contents in garnet in prograde metamorphism follow a similar pattern to that of MnO contents. All the individual garnets show a similar zonation from a more calcic core to a less calcic rim and there is a general trend of decreasing CaO content with increasing metamorphuic grade. In the previous example of Saint-Ponais (Montagne Noire), the core of the garnets at the +garnet isograd, have a content in CaO of about 6.5–7 wt%

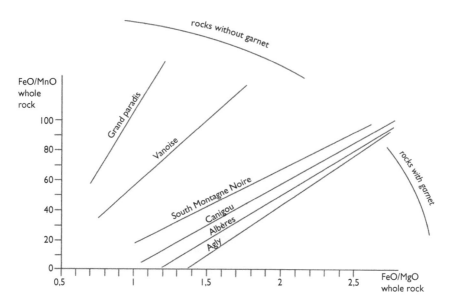

Figure 3.33 Diagram FeO/MgO vs FeO/MnO in whole rock – comparison of different types of metamorphism *(data by Fonteilles and Guitard, 1971, Chopin, 1979, Demange, 1982).*

and about 4 wt% at their rim; garnets in the sillimanite + muscovite zone have CaO contents of 2–1.5%. The anorthite content of plagioclase grows correlatively.

High grade metamorphism of pelitic rocks

At high temperatures, muscovite is destabilized into sillimanite + K-feldspar + water. This is often called the "second sillimanite isograd", the first one being the polymorphic transformation andalusite = > sillimanite) (Figure 1.9).

At higher temperatures, aluminous biotites is gradually destabilized by a continuous reaction, into garnet + K-feldspar + water. Garnet and sillimanite therefore characterized paragneiss of pelitic origin at high grade metamorphism. Kinzigite are paraagneisses containing biotite–sillimanite–garnet–cordierite–quartz–feldspar–plagioclase (and graphite) that are derived from peraluminous shales. Prograde metamorphism of metapelitic rocks is no more emphasized by isograds but by gradual changes in composition of minerals. A very high grade, in granulite facies, even biotite finally disappears. Spinel (+quartz) or sapphirine (+quartz) can develop at the expense of cordierite.

Sapphirine is a mineral of high grade metamorphism, upper amphibolite facies and granulite facies. It only appears in the aluminum and magnesium-rich and silica-poor rocks. The compositions of such rocks are very different from any ordinary magmatic or sedimentary rocks. Sapphirine is associated with minerals such as cordierite, magnesian garnet, corundum, kyanite or sillimanite, phlogopite and spinel.

Metamorphism of alterites, paloesols and bauxites

Supergen alteration leaches the alkalis, calcium, magnesium and possibly silica, producing rocks(/soils) rich in aluminum and iron: ferrallitic soils and bauxite.

The andalusite–chloritoid–chlorite–muscovite–quartz schists, in domains of weak metamorphism in the Pyrenees and the Montagne Noire and some kyanite–muscovite "quartzites" (of central Spain) probably derive from ferrallitic paleosols more or less reworked.

Metamorphism of bauxites produces emery deposits with corundum, kyanite, staurolite, margarite. Diaspore and pyrophyllite occur at a lower metamorphic grade.

Metamorphism of iron formations

Almandine (with some proportion of spessartine) appears in the metamorphosed iron formations in amphibolite and granulite facies.

High pressure metamorphism of basic rocks

Plagioclase is unstable at high pressure. The associations plagioclase + calcic pyroxene or plagioclase + hornblende characteristic of basic rocks, is then replaced by an association of sodic pyroxene or amphibole + an aluminous phase. This phase is commonly an aluminous garnet. If the primary plagioclase was an anorthite-rich one (i.e. rich in aluminum), there may be formation of kyanite.

Eclogites are rocks rich in magnesian garnet + sodium-rich pyroxene. As well as the composition of pyroxene varies depending on the type of eclogite, the composition of the garnet varies: eclogites richer in almandine in relatively low temperatures are associated with domains of blueschist facies; high-temperature eclogites are richer in pyrope. Some lower temperature eclogites contain kyanite and/or magnesian staurolite (the latter is stable only in the absence of quartz); some higher temperatures eclogites contain sapphirine.

Amphibolites of medium to high pressure may contain garnet or, more rarely, kyanite. Note that the presence of garnet in an amphibolite is not a sufficient criterion to say that it is a rock of medium to high pressure. Garnet occurs in low pressure amphibolites when the Fe/Mg ratio of these rocks is high enough.

3.3.2.2 Igneous rocks

Peraluminous igneous rocks:

*   either belong to peraluminous suites (for example, the Laouzas granite, Montagne Noire, France)
*   or are the differenciated members of mainly metaaluminous suite (like the ordinary calcalkaline series) such as evolved granites, aplites and pegmatites.

The peraluminous character is expressed by the presence of minerals such as muscovite, garnet, cordierite, aluminous biotite, sometimes topaz and/or beryl, seldom andalusite and, exceptionally, kyanite.

The garnets from differentiated granitoids, aplites and pegmatites belong to the almandine–spessatine series.

The habit of cordierite may be either euhedral crystals, clearly of magmatic origin, or nodules ("chestnuts") with inclusions of sillimanite and biotite, probably metasomatic origin. Both habits sometimes may occur simultaneously in a same sample:

Cordierite and topaz are also found in some rare rhyolites.

Kimberlites, peridotites, biotite peridotites contain a pyrope-rich garnet containing a greater or lesser proportion of chromium.

3.3.2.3 Metasomatic rocks

Greisens are rocks made of quartz and muscovite formed by leaching of alkalis from quartzo-feldspathic rocks. Greisens frequently contain topaz and beryl, accompanying mineralization in tin and/or tungsten.

Further leaching of the alkalis leads to "secondary hydrothermal quartzites", rocks formed mostly of quartz with aluminous minerals: andalusite, corundum, pyrophyllite, diaspore, etc. These rocks develop at the expense of granitoids, volcanic rocks and, more rarely, sedimentary rocks in an environment of volcanic or subvolcanic rocks of acid to intermediate composition.

The quartz + pink andalusite cores of pegmatites come from similar leaching.

This is probably the same phenomenon that produces quartz + pink andalusite lenses or quartz + blue kyanite lenses (with some staurolite), that are frequent in schists and quartz + sillimanite nodules and lenses in gneisses.

Corundum occurs in pegmatites associated with granitoids or nepheline syenites. The presence of corundum implies leaching of alkalis and above all the absence of quartz. Nepheline syenite pegmatites contain no quartz. It is probable that the granitic pegmatites have undergone desilication in contact with limestone.

Emerald occurs in biotitites derived ultramafic rocks by potassic metasomatism due to fluids, probably of perigranitic (/pneumatolitic) origin.

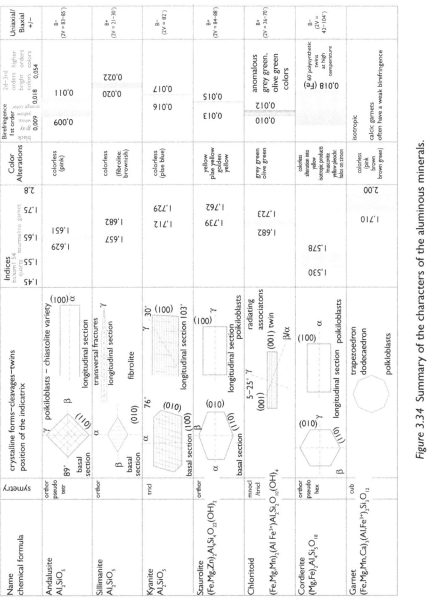

Figure 3.34 Summary of the characters of the aluminous minerals.

3.3.2.4 Sedimentary rocks and alterites

Andalusite, kyanite, staurolite, garnet, topaz and corundum are rather inalterable and of high density minerals.They are concentrated in the clastic alluvial and beach (intertidal) sediments (black sands). Placers are also the main source of rubies and sapphires.

Diaspore, boehmite and gibbsite are the main constituents of bauxites.

3.4 CALCIC, MAGNESIAN AND CALC-MAGNESIAN MINERALS

These calcic, magnesian and calc-magnesian minerals are the major minerals of calcic and magnesian carbonate rocks, metamorphic rocks (impure marble, calc-silicate-gneiss) and metasomatic rocks (skarns). These minerals also occur in calcic and magnesian basic metamorphic rocks of igneous origin. Some are rare primary minerals of alkaline igneous rocks. The traditional mineralogical classification ranges the various minerals discussed here in various classes. The similarities in chemical composition and the fact that these minerals are found associated in their occurrences, justify their regrouping here.

3.4.1 Chemical composition and stability

These minerals are here classified according to their chemical composition (Table 3.2 and Figure 3.35):

* carbonates,
* calcic aluminous silicates,
* calcic non-aluminous silicates,
* non-aluminous magnesian and calcic-magnesian silicates,
* calcic ferro-magnesians silicates,
* minerals of calcium and titanium: titanite and perovskite (in part 3–5).

This classification is somewhat arbitrary. Indeed, some minerals or mineral groups, such as calcic garnets or melilite, can be classified in several of these groups. Nevertheless, this classification allows us to link the composition of these minerals with the one their host rock and thus their occurrences.

3.4.1.1 Carbonates

Common carbonates belong to two groups: the trigonal carbonates and orthorhombic carbonates.

Table 3.2 Calcic, magnesian and calc-magnesian minerals.

carbonates

calcite	$CaCO_3$
dolomite	$CaMg(CO_3)_2$
magnesite	$MgCO_3$
aragonite	$CaCO_3$

anhydrous Ca–Al minerals, without Fe–Mg

anorthite	$CaAl_2Si_2O_8$
Tschermak molecule	$CaAlAlSiO_6$
grossular	$Ca_3 Al_2Si_3O_{12}$
gehlenite	$Ca_2 Al_2SiO_7$
Na-melilite	$NaCaAlSi_2O_7$

Ca–Si minerals without sans Al, nor Fe–Mg, nor volatiles

larnite	Ca_2SiO_4
rankinite	$Ca_3Si_2O_7$

Ca–Si minerals with volatiles

tilleyite	$Ca_3Si_2O_7 \cdot CaCO_3$
spurrite	$2(Ca_2SiO_4) \cdot CaCO_3$

Ca Al minerals without Fe–Mg, with water, volatiles

prehnite	$CaAl_2Si_3O_{10}(OH)_2$
zoisite	$Ca_2 Al Al_2Si_3O_{12}(OH)$
lawsonite	$CaAl_2Si_2O_7(OH)_2, H_2O$
meionite	$3(CaAl_2Si_2O_8), Ca(CO_3, SO_4)$
marialite	$3(NaAlSi_3O_8), NaCl$

Ca Fe–Mg minerals without Al nor water

wollastonite	$CaSiO_3$
diopside	$CaMgSi_2O_6$
hedenbergite	$CaFeSi2O_6$
andradite	$Ca_3(Fe^{3+},Ti)_2Si_3O_{12}$
periclase	MgO
forsterite	Mg_2SiO_4
åkermanite	$Ca_2 MgSi_2O_7$
monticellite	$Ca\ Mg\ SiO_4$

Ca Fe, Mg minerals, with water – volatiles, without Al

tremolite	$Ca_2(Fe, Mg)_5Si_8 O_{22}(OH)_2$
talc	$Mg_6(Si_8O_{20})(OH)_4$
brucite	$Mg(OH)_2$
Norbergite	$Mg_2SiO_4 \cdot Mg(OH, F)_2$
Chondrodite	$2\ Mg_2SiO_4 \cdot Mg(OH, F)_2$
Humite	$3\ Mg_2SiO_4 . Mg(OH, F)_2$
clinohumite	$4\ Mg_2SiO_4 \cdot Mg(OH, F)_2$

(Continued)

Table 3.2 (Continued)

Aluminous, hydrated Ca Fe, Mg minerals	
hornblende	$(Na, K)_{0-1}Ca_2(Fe, Mg, Al)_5(Si, Al)_8 O_{22}(OH)_2$
Common hornblende	$Ca_2(Mg, Fe)_4 Al_2Si_7O_{22}(OH)_2$
pargasite	$NaCa_2 Mg_4 Al_3Si_6O_{22}(OH)_2$
pistachite	$Ca_2(Al, Fe^{3+})Al_2Si_3O_{12}(OH)$
pumpellyite	$Ca_4(Mg, Fe^{2+})(Al, Fe^{3+})_5O(OH)_3(Si_2O_7)_2(SiO_4)_2$ $2 H_2O$
vésuvianite	$Ca_{10}(Fe, Mg)_2 Al_4(Si_2O_7)_2(SiO_4)_5(OH, F)_4$
esseneïte	$CaAlFeSiO_6$
merwinite	$Ca_3 MgSi_2O_6$

Trigonal carbonates form three series:

- calcite $CaCO_3$
  Substitution of Ca^{++} by other divalent ions is very limited due to the difference in ionic radius between these ions. At equilibrium, the molar percentage of $MgCO_3$ in calcite at ordinary temperatures does not exceed a few %. It may be higher, in metastable state, in some sedimentary conditions, or in some organisms. This magnesium-rich calcite recrystallizes into magnesian calcite + dolomite. The term magnesian calcite is used when this percentage exceeds 4%. This percentage of $MgCO_3$ in calcite in equilibrium with dolomite increases with temperature reaching 20% around 800°C. Iron, manganese and, to a lesser extent, strontium also enter in limited quantities in the lattice of calcite;
- dolomite $CaMg(CO_3)_2$ – ankerite $Ca(Fe, Mn)(CO_3)_2$ series
  Dolomite has a structure similar to the one of calcite but half the Ca ions are replaced in an ordered manner by Mg ions. There is a continuous solid solution between the $CaMg(CO_3)_2$ end member and the $CaFe(CO_3)_2$ end member. The word dolomite is restricted to minerals where the ratio Mg/Fe is higher than 4. It may come up about 3 wt% MnO in the lattice:

  –magnesite $MgCO_3$–siderite $FeCO_3$–rhodocrosite $MnCO_3$ continuous series

Although the various trigonal carbonates have some differences in color and indexes (siderite may be brownish in thin section), the most convenient way to distinguish these minerals in thin section is staining: the attack by a weak acid solution of alizarin distinguishes calcite that is stained in pink to

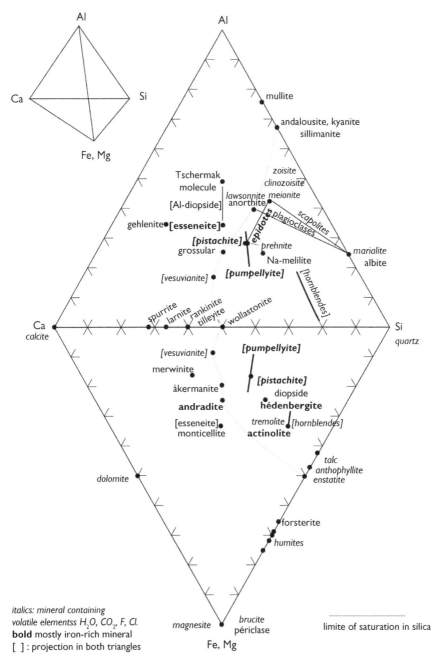

Figure 3.35 Composition calcic, magnesian and calc-magnesian minerals in the diagramme Ca–Si–Al–Fe–Mg diagram. Projection of two faces of this tetrahedron: some minerals are actually in the Ca–Al–Si or Ca–Si–Fe, Mg planes; others at the center of the tetrahedron project on these two planes.

red, from dolomite, which remains colorless; potassium ferrocyanide distinguishes iron minerals: ankerite is colored in blue, siderite in black.

Aragonite is the only common orthorhombic carbonate in the rocks. It is a high pressure polymorph of calcite but can exist in the metastable state at low pressure and at ordinary temperature under specific conditions: shells of mollusks, some hydrothermal deposits and sedimentary environments. Aragonite, either formed in metastable state or from high-pressure metamorphism, is more or less rapidly transformed into calcite.

3.4.1.2 Aluminous, anhydrous and hydrated, calcic silicates

These minerals have compositions close to that of anorthite (Table 3.2) and play a similar role in conditions where it is unstable (Figure 3.4 and §3.1.2.2).

At low temperatures, the hydrated Ca–Al minerals (epidote, prehnite, lawsonite) play the role of anorthite under conditions of low temperature metamorphism at low or high pressure. Scapolite plays the role of feldspar in the presence of volatile elements (NaCl, CO_3, SO_4). Gehlenite of the melilite group plays the role of a calcic plagioclase in rocks under-saturated in silica.

Epidote, scapolite, calcic garnet and melilite calcium form families larger than the above Ca–Al end members.

Epidote group

Epidotes are sorosilicates made of chains of octahedra sharing an edge in b direction (M site). These chains are linked by SiO_4 tetrahedra and Si_2O_7 double tetrahedra. Y octahedral sites are attached to the chains. There are also between the chains the 9-fold and 10-fold coordinated sites occupied by large cations (X site). The general formula of epidote is:

X_2 Y M_2 SiO_4 Si_2O_7.O OH
X = Ca, Ce, La, Y, Th, Fe^{2+}, Mn^{2+}, Mn^{3+}
Y = Al, Fe^{3+}, Mn^{3+}, Fe^{2+}, Mn^{2+}, Ti
M = Al, Fe^{3+}, Mn^{3+}

Epidotes *s. s.* form a series of general formula $Ca_2(Al, Fe^{3+})Al_2Si_3O_{12}(OH)$ between a purely aluminous end member $Ca_2Al Al_2Si_3O_{12}(OH)$ and a ferric-iron $Ca_2Fe^{3+} Al_2Si_3O_{12}(OH)$. The iron-bearing end member, *pistachite*, is monoclinic. The aluminous end member has two polymorphs: a monoclinic form of low-pressure, *clinozoisite* and an orthorhombic high pressure form *zoisite*. Fe_2O_3 content of zoisite does not exceed 2 wt%. Thulite is a manganesiferous variety of zoisite containing up to 0.5 wt% of MnO.

Piemontite is a manganesiferous epidote where manganese is in the Mn^{3+} form: $Ca_2(Mn^{3+}, Fe^{3+}, Al) Al_2Si_3O_{12}(OH)$. Piemontite form a continuous series from pistachites to an end member of formula $Ca_2Mn^{3+}Al_2Si_3O_{12}(OH)$. Aluminum in M position M may be partially replaced by Mn^{3+}, defining an end member of formula $Ca_2Mn^{3+}_2AlSi_3O_{12}(OH)$.

Rare earth elements (REE) enter in the lattice of epidote by substitution:

$$Ca^{2+}\ Fe^{3+} \Leftrightarrow REE^{3+}\ Fe^{2+}$$

In addition, Th, U, Mn^{2+} may enter in the X site and the beryllium may enter as a substitution of Si. *Allanite* (also known as *orthite*) is a REE epidote of formula:

$$(Ca, Ce, La, Y, Th, U, Mn^{2+})_2(Al, Fe^{3+}, Fe^{2+})Al_2Si_3O_{12}(OH)$$

Pumpellyite has a structure similar to that of epidote: chains of octahedra, linked by SiO_4 tetrahedra and Si_2O_7 double tetrahedra. This gives place to 7-fold coordinated sites, occupied by Ca in the chains of octahedra, four sites are occupied by only Al; two sites contain (Al, Fe, Mg). OH ions partially replace oxygen. Hence the formula is:

$$Ca_2(Al, Mg, Fe^{2+}, Fe^{3+})Al_2(Si_2(O, OH)_7) (SiO_4) (O, OH)_3$$

The composition of pumpellyite varies by substitutions $Al \Leftrightarrow Fe^{3+}$ and $Mg \Leftrightarrow Fe^{2+}$ and, rarely, by the presence of chromium and manganese.

The stability range is limited by temperature: the pumpellyite is replaced at lower temperatures (below 200–250°C) by the association chlorite + prehnite and towards higher temperatures (above 350–400°C) by zoisite + chlorite + calcic garnet. These reactions define the prehnite–pumpellyite facies between zeolite and greenschist facies. Pumpellyite is also found at higher pressures in blueschist facies.

Lawsonite is a mineral close to epidote in its structure: chains of aluminum octahedra, linked by Si_2O_7 groups This gives place for sites for Ca and OH. Chemically lawsonite varies little: $CaAl_2Si_2O_7(OH)_2,H_2O$; this formula is similar to that of anorthite except for the presence of H_2O and a more compact structure that makes lawsonite a mineral of high-pressure. At lower pressures and higher temperatures it is replaced by zoisite + kyanite + quartz (Figures 1.9 and 3.4) and at "low" pressure (up to 10kb) and at high temperature, by anorthite.

Prehnite

Prehnite $Ca_2Al_2Si_3O_{10}(OH)_2$ is a phyllosilicate, made of Si_3AlO_{10} sheets, where alumina is in the tetrahedral position, bound by chains Al(in octahedral position)–Ca (7-fold coordination)–OH. These sheets determine the perfect (001) cleavage. The composition of prehnite varies little, the only significant substitution is that of octahedral alumina by ferric iron.

Prehnite is a mineral of low temperature (less than 380–400°C), characteristic of zeolites and prehnite–pumpellyite facies.

Scapolite group

Scapolite are tectosilicates forming a series between the marialite end member $3(NaAlSi_3O_8)$, NaCl and meionite end member $3(CaAl_2Si_2O_8)$, $Ca(CO_3, SO_4)$. Aluminum is in tetrahedral sites. These are the equivalents of plagioclase but with a looser structure permitting the introduction of large anions Cl^-, CO_3^{2-}, SO_4^{2-}.

Terms close to marialite are sometimes called dipyre. Intermediate terms were called wernerites.

Scapolites are minerals of medium (550–600°C) at high (850–1050°C) temperature. The meionite end member is stable at only temperatures above 800°C. They are also stable at relatively high pressures (up to 20 kb). Their stability depends, of course, on the fluid fugacity and especially that of CO_2.

Calcic garnets

Calcci garnets are characterized by calcium as the main cation in X site (§3.3.1.2). The three end members («ougrandites») are:

Ouvarovite $Ca_3Cr_2Si_3O_{12}$
Grossular $Ca_3 Al_2Si_3O_{12}$
Andradite $Ca_3Fe^{3+}_2Si_3O_{12}$

Ouvarovite is a rare garnet limited to the chromite deposits associated with ophiolitic complexes or with mafic–ultramafic complexes of Bushveld type.

Garnets of the calcic metamophic rocks (marbles, calc–silicate–gneiss, skarns) belong to the **grossular–andradite** series ("grandites").

Andradite from alkaline igneous rocks frequently contain titanium, $Ca_3(Fe^{3+}, Fe^{2+}, Ti)_2Si_3O_{12.}$ This element enters the lattice in Y site by substitution:

$$2Fe^{3+} \Leftrightarrow Fe^{2+} Ti$$

Some authors consider that titanium may also enter into the tetrahedral site as a substitution to silica. Titaniferous garnets are called melanite if $Fe^{3+} > Ti$ and schorlomite $Fe^{3+} < Ti$.

Zirconium associated with titanium, can enter into the octahedral site in andraditic garnet. There is then substitution of tetrahedral silica Al and/ or Fe^{3+}:

$$Al \, Si \Leftrightarrow Zr \, (Al, \, Fe^{3+})$$

Such zirconian garnet is called kimzeyite: $Ca_3(Zr, Ti)_2(Si, Al)_3O_{12}$.

Garnet of some skarns may contain some few percent of Sn. This element is probably in tetrahedral site in substitution to silica. He may enter into octahedral Y site by substitution $Fe^{3+} \, Si \Leftrightarrow Sn \, Al^{3+}$ and/or in X site as Sn^{2+}.

Vanadium can enter in the octahedral site replacing Al (goldmanite end member: $Ca_3V_2Si_3O_{12}$).

Hydrogrossular are grossular in which part of the SiO_4^{4-} tetrahedra is replaced by OH; their formula is thus $Ca_3Al_2Si_2O_8(SiO_4)_{1-x}(OH)_{4x}$. The iron content generally remains very limited.

Vesuvianite

Vesuvianite (also called idocrase) is a mineral whose structure and chemical composition are similar to those of hydrogrossular. It differs from it by the addition of atoms of aluminum, iron, and magnesium along an axis of symmetry of order 4, which gives to it a tetragonal symmetry. Silica occupies 10 tetrahedral SiO_4 sites as in the garnets, and also 8 Si_2O_7 double tetrahedra sites; vesuvianite is thus classified in the sorosilicates.

The main chemical difference with hydrogrossular is the presence of magnesium, an element that probably stabilizes vesuvianite in its paragenesis.

The general formula of vesuvianite is:

$$Ca_{19}(Fe, Mg)_3(Al, Fe)_{10}(Si_2O_7)_4(SiO_4)_{10}(OH, F)_{10}$$

Composition of vesuvianite varies considerably by the proportions of iron, magnesium and aluminum. Calcium can be replaced by small amounts of Na, K, Li, Mn, U, Th and LREE; iron and aluminum may be replaced by Cr, and Ti; silica by Be.

Vesuvianite is a mineral of medium to high temperature (360–650/700°C). Its stablility field strongly depends on the fluid pressure and the nature of the fluid. At higher temperatures it is replaced by grossular–monticellite and wollastonite.

Melilite group

Melilite group minerals have a structure in sheets consisting of octahedral Y sites forming a square centered pavement (an octahedron at each corner and one in the center) linked by double tetrahedra Si_2O_7 (Z sites). These sheets are linked by Ca ions (X sites), cleavage (001) of the melilites corresponds to that sheet structure. The tetragonal symmetry is due to the square centered pavement. The general formula of melilite is:

X_2 Y Z_2 O_7
X = Ca, Na, (K, Sr)
Y = Al, Mg, (Fe^{2+}, Mn)
Z = Si, Al

In *åkermanite* Ca_2MgSi_2O, Y sites are occupied by Mg. The formula of *gehlenite* $Ca_2Al_2SiO_7$ is deduced from it by the substitution Mg Si ⇔ Al^{IV} Al^{VI}. The formula of **soda melilite** (Na-melilite, or melilite) $NaCaAlSi_2O_7$ is deduced by the substitution Ca Mg ⇔ Na Al. There are certainly gaps of solubility between these end members, but they remain poorly investigated.

Åkermanite and gehlenite are minerals under-saturated in silica. They are only very rarely associated with plagioclase.

Minerals of the melilite group are high temperature minerals over 700°C for åkermanite (this mineral is characteristic of the sanidinite facies), 600°C for gehlenite.

3.4.1.3   Non aluminous calcium silicates

The only relatively common, calcium silicate is *wollastonite* $CaSiO_3$.

Wollastonite is an inosilicate, the chains of SiO_4^{4-} tetrahedra differ from those of pyroxenes by the fact that the pattern is formed by two tetrahedra pointing in one direction alternating with a tetrahedron pointing in the opposite direction (Figure 1.8).

There are three polymorphs: triclinic pseudowollastonite of very high temperature (above 1120°C), monoclinic wollastonite-2M and triclinic wollastonite-Tc at lower temperatures. The usual form of wollastonite is the triclinic form.

Wollastonite accepts small quantities of iron and manganese in substitution to calcium.

Wollastonite is stable at high temperatures. The reaction:

calcite + quartz ⇔ wollastonite + CO_2

occurs around 650–750°C, depending on the CO_2 pressure. Towards higher temperatures wollastonite is transformed into spurrite in the presence of calcite at about 1000°C.

There are many other calcium silicates, most of them are anhydrous but they may contain anions such as CO_3: for instance, *larnite* (Ca_2SiO_4), *rankinite* ($Ca_3Si_2O_7$), *spurrite* ($2Ca_2SiO_4 \cdot CaCO_3$), *tilleyite* ($Ca_3Si_2O_7 \cdot CaCO_3$). These are rare minerals, limited to high temperature skarns. Larnite and spurrite are presented in the CD.

Pectolite $NaCa_2Si_3O_8(OH)_2$ (triclinic) is an inosilicate with a structure similar to that of wollastonite. Mn may partially replace calcium, Mg, Fe as substitution to calcium, Al as substitution to Si may also be present. It has been synthetized at a temperature of 180°. So it is mostly a hydrothermal mineral (but may also occur as a primary mineral in skarns and rare alkaline igneous rocks).

Charoite $K_5Ca_8(Si_6O_{15})_2(Si_2O_7)Si_4O_9(OH) \cdot 3H_2O$ $(=K(Ca, Na)_2Si_4O_{10}$ $(OH, F) \cdot H_2O)$ (monoclinic) is a very rare inosilicate, known only in the Murunskii massif (Yakutia, Russia). It has a quite unique structure of a 3-periodic single and multiple chains. The only justification to show it in the CD is its strange beauty.

3.4.1.4 Magnesian, non-aluminous silicates (oxides and hydroxides) (anhydrous and hydrated; saturated and under-saturated in silica)

These are minerals of metamorphism of siliceous dolomitic rocks and of skarns developed on dolostones. Monticellite and humites are also of primary minerals of alkaline ultramafic rocks. Talc and brucite are also products of alteration of olivine and serpentine.

This group includes:

- peridots and minerals close to them: forsterite Mg_2SiO_4, monticellite $CaMgSiO_4$, humites group(1–4) $Mg_2SiO_4 \cdot Mg(OH,F)_2$;
- diopside $CaMgSi_2O_6$;
- merwinite $Ca_3MgSi_2O_6$, very rare mineral of high temperature (not discussed here);
- talc, hydrated phyllosilicate $Mg_6(Si_8O_{20})(OH)_4$;
- periclase MgO and brucite $Mg(OH)_2$.

Clintonite and xantophyllite $Ca_2(Mg, Al)_6Si_{1.5}Al_{5.5}O_{20}(OH)_4$, are magnesian (and aluminous) brittle micas (§3.2.1.7) which have the same occurrences as the previous minerals.

Neither enstatite $MgSiO_3$ nor anthophyllite $Mg_5Si_8O_{22}(OH)_2$ occur in this type of deposits because they are replaced in presence of calcite by forsterite + diopside.

$$3MgSiO_3 + CaCO_3 \Leftrightarrow CaMgSi_2O_6 + Mg_2SiO_4 + CO_2$$

Forsterite and diopside have been previously discussed with olivines and pyroxenes.

The composition of *monticellite* $CaMgSiO_4$ does not vary very much as the substitution of Mg by Fe or Mn is very limited. Monticellite is a mineral of high temperature, formed at temperatures higher than diopside, forsterite and wollastonite (650–850°C depending f_{CO2}). At higher temperatures, it is replaced by åkermanite (wollastonite + monticellite ⇨ åkermanite at 700°C), spurrite, merwinite.

Humites group includes four minerals which are very similar in structure, chemical composition, optical properties and occurrences:

Norbergite	$Mg_2SiO_4 \cdot Mg(OH, F)_2$	orthorhombic
Chondrodite	$2Mg_2SiO_4 \cdot Mg(OH, F)_2$	monoclinic
Humite	$3Mg_2SiO_4 \cdot Mg(OH, F)_2$	orthorhombic
Clinohumite	$4Mg_2SiO_4 \cdot Mg(OH, F)_2$	monoclinic

Fe, Mn, Zn may replace Mg in very small quantities. Titanium (Ti^{4+}) may enter as a substitution to magnesium (Mg^{2+}) in $Mg(OH)_2$ groups, the electrical neutrality being maintained by the substitution of OH^- by O^{2-}.

Humites are medium to high temperature minerals. The association humite + calcite is formed at the expense of dolomite + forsterite/tremolite or talc associations. The stability of these associations depends on the relative pressures of CO_2 and H_2O. Their appearance, of course, depends of fluorine.

Talc $Mg_6(Si_8O_{20})(OH)_4$ is a phyllosilicate whose structure is similar to that of the TOT sheets of trioctahedral micas: two layers of SiO_4^{4-} tetrahedra sandwiching an octahedral layer which sites are occupied by Mg. Unlike mica there is no interlayer site. The connection between the layers is a very weak bonding of the Van der Waals type. This explains the cleavage of talc, its very low hardness (1 by definition in the Mohs scale) and greasy touch. The substitutions Mg ⇔ Fe and Mg Si ⇔ $Al^{IV} Al^{VI}$ are very limited.

Talc is a mineral of low to medium temperature. Its stability towards higher temperatures, controlled by the water pressure is limited by the appearance of anthophyllite and enstatite.

Periclase MgO (cubic mineral, magnesium may be replaced by Fe in very small quantities) is a mineral from high temperature metamorphism of dolostones. It is formed at higher temperatures than wollastonite, but lower than monticellite.

Periclase is generally destabilized into **brucite** $Mg(OH)_2$ (with very little substitution of Mg by Fe), a fibrous mineral, trigonal, with a layered structure. Brucite is altered into hydromagnesite $3MgCO_3 Mg(OH)_2 3H_2O$.

3.4.1.5 *Ferro-magnesian calcic silicates*

These minerals share the same occurrences as other minerals presented in this chapter; have been treated previously. The most important are amphibole and pyroxene:

– **Calcic amphiboles**

– non aluminous *tremolite–actinolite–*ferroactinolite series

$Ca_2(Fe, Mg)_5Si_8 O_{22}(OH)_2$

- aluminous: *hornblendes*

$(Na, K)_{0-1}Ca_2(Fe, Mg, Al)_5(Si, Al)_8 O_{22}(OH)_2$

Among the hornblendes, pargasite $NaCa_2 Mg_4Si_6 Al_2O_{22}(OH)_2$ occurs in the metamorphic dolomitic limestones and dolostones; its ferro-magnesian equivalent, hastingsite occurs in skarns.

- Pyroxenes
  - *diopside* $CaMgSi_2O_6$–*hedenbergite* $CaFe^{2+}Si_2O_6$–*johannsenite* $CaMnSi_2O_6$ series
  - *aluminous diopsides*

3.4.2 Occurrences

3.4.2.1 *Sedimentary rocks*

Calcite and dolomite are the major constituents of sedimentary carbonate rocks: limestones and dolostones, and carbonate rocks containing a greater or lesser proportion of clay, quartz, etc. (marls, calcareous sandstone, etc. Figure 1.10).

Minerals that precipitate from seawater are magnesian calcite and aragonite. Aragonite may form ooliths and carbonate muds; it is associated with gypsum and celestine in evaporitic environment; it also may form the cement of intertidal deposits.

Calcite may also be formed through living organisms (protists, stromatolites, algae, sponges, corals, echinoderms, etc.). Aragonite also forms the shell or skeleton of some animals (mollusks, and some corals). Bioclast thus form an important part of the carbonate deposits.

Since aragonite is metastable and recrystallizes into calcite, and magnesian calcite loses its Mg, most carbonate sediments are low-Mg calcite.

Calcite and aragonite form also karstic cave formations (stalagmites, stalactites, etc.) Aragonite, in particular, forms cave pearls, cave flowers and frostwork.

Dolomite may precipitate directly from seawater in so far that its Mg/Ca ratio is increased: supersaturated anaerobic lagoons and deep seated sediments rich in organic material. In a lagoonal environment, the order of precipitation follows the order of increasing solubility in seawater: calcite, dolomite, gypsum/anhydrite, halite, potassium and magnesium salts.

Most dolomites do not seem to have been formed in such extreme environments. The process is probably a diagenetic evolution caused mixing of freshwater and saltwater by subsurface conditions. This mixing lowers the salinity of seawater and/or allows the dissolution of magnesium either from magnesian calcite either from clay, which increases the Mg/Ca ration of the waters. Sub-saturation of these waters in calcite and supersaturation in dolomite occurs then: dolomite precipitates and therefore replaces calcite.

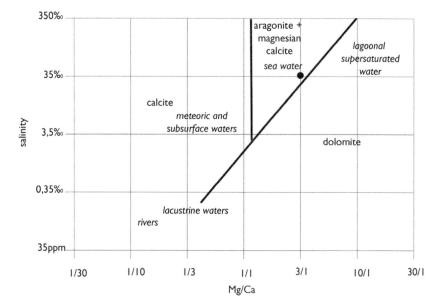

Figure 3.36 Field of precipitation of carbonates: calcite, dolomite, aragonite, magnesian calcite (after Folk and Land, 1975).

Secondary dolomites are the product of a metasomatic transformation of limestone. These transformations may occur a long time after the primary deposit. The source of magnesium responsible for these secondary dolomitization generally remains unknown.

3.4.2.2 Metamorphic rocks

Metamorphism of basic rocks

Many facies of metamorphism (Figure 1.9) are defined by associations resulting from the metamorphism of mafic rocks of basaltic composition (calcic plagioclase + pyroxene). The anorthite is unstable at low temperature and high pressure, its role is played in these rocks by various calcium silicates.

In metamorphism of low to medium pressure, the following sequence is observed:

- in zeolites facies: heulandite/clinoptinolite $(Ca, Na)_2 Al_2Si_7O_{18} . 6H_2O$), then laumontite $CaAl_2Si_4O_{12} . 4H_2O$;
- prehnite and pumpellyite in prehnite–pumpellyite facies;
- epidote in greenschist then in epidote–amphibolite facies.

At higher pressure, the presence of zoisite or lawsonite in the domains of higher P/T gradient (Figure 1.9) divides the blueschist facies into two sub-facies.

Metamorphism of calcic, dolomitic rocks and skarns

Dolomite is a mineral that becomes unstable very rapidly in the presence of silicates as the metamorphism becomes significant. Prograde metamorphism of *dolostones* or *dolomitic limestones* siliceous is marked by a fairly constant succession of isograds:

- dolomite + quartz paragenesis ($+H_2O$) is transformed into talc + calcite + CO_2 from the greenschist facies up; if water pressure is too low, tremolite is the first mineral to appear, instead of talc;
- the association tremolite + calcite occurs in carbonate rocks roughly simultaneously with that of aluminous silicates in metapelites;
- then comes the diopside isograd + calcite;
- forsterite (+calcite) (and humite) appear at a high grade metamorphism, beyond the isograd disappearance of muscovite in meta-pelites;
- periclase is exceptional in the regional metamorphism.

Impure marbles and calc–silicate–gneisses are rocks of extremely varied chemical composition (Figure 1.10) by the relative proportions of carbonate (calcite and/or dolomite), silica and clay minerals and by variations in the composition of clay minerals (Si/Al and Fe/Mg). These variations may occur at very low scale. The presence of halite in the initial sediments may be responsible for the appearance of scapolite ("couseranite" a local name dipyre/marialite in the Pyrenean metamorphism). Phenomena of diffusion between carbonate beds and pelitic beds (reaction-skarn) during metamorphism are superimposed to the initial variability of the sediments: such diffusion phenomena may be (partly) responsible for the banding of the calc–silicate–gneisses.

The study of the metamorphism of these rocks is not only complicated by the chemical complexity, but also by the fact that the metamorphism of rocks depends not only on temperature but also on the partial pressures of H_2O and CO_2.

It is therefore impossible to present a general picture for the prograde metamorphism of such rocks. Their interest in the study of regional metamorphism remains modest.

Skarns are metasomatic rocks developed at the contact between two chemically incompatible media, most often between a carbonate rock, limestone or dolostone (but also basic rocks) and siliceous rock (or magma). This siliceous rock is often an igneous rock of the family of granitoids. The example of enclaves in the lava of Vesuvius and the famous Scawt Hill skarns (Tilley and Harwood, 1931) also display skarns developed at contact with basic igneous rocks. The skarns are formed mainly of anhydrous or water-poor silicates of calcium, magnesium, iron and manganese.

Endoskarns are skarns formed at the expense of igneous or siliceous rocks; *exoskarns* are those formed at the expense of carbonate rocks.

Essentially, there is a transfer of silica and iron from the siliceous rock to the calcic rock and transfer of calcium from the calcic rock to siliceous rock: Ca, Si, Fe are mobile elements. A metasomatic column (which can reach several tens of meters of thickness) is so formed; it is made of zones with sharp boundaries and contrasting mineralogy. The activity of silica increases from the calcic rock to the siliceous rock and the activity of calcium varies inversely.

Aluminum, magnesium, zirconium (and to a lesser extent, titanium, phosphorus, rare earths) are inert elements or of very little mobility in this process. The presence of minerals containing characteristic elements of igneous rocks (which do not occur in carbonate rocks), such as Al, Zr, P, allows us to precisely define the boundary between endo-and exoskarns.

Two main types of skarns may be distinguished by the nature of the carbonate rock: calcic skarns developed on limestone and magnesian skarns developed on dolostones. There are also very rare manganese skarns developed on manganiferous rocks. Skarns that grow on calc–silicate–gneisses and calcium are called skarnoides or calcic hornfels.

The evolution of skarns can generally be described in several stages:

At the *magmatic stage* (usually there are several sub-stages), the exchanges occur between the magma and the host rock. The magma is contaminated by the input of calcium and crystallizes aluminous diopside. It may even become under-saturated in silica; this is reflected by the crystallization of melilite of the åkermanite – Na-melilite series and of feldspathoids. Minerals developed in the marbles are anhydrous minerals high temperature (700–900°C): wollastonite, tilleyite, spurrite, rankinite, larnite in calcic skarns, monticellite, forsterite, åkermanite, merwinite, periclase in magnesian skarns. This stage is very rarely preserved in calcic skarns; it is somewhat better preseved in magnesian skarns.

The *main stage* (650–400°C), commonly observed, is a hydrothermal stage where mainly anhydrous minerals are formed: garnets of the grossular–andradite series, pyroxene of the diopside–hedenbergite–johannsenite series, wollastonite, calcic plagioclase, scapolite, vesuvianite and, in magnesian skarns, forsterite, humites, phlogopite, hastingiste, etc. Endoskarns usually contain hydrated minerals such as amphibole of the hornblende group. This stage is often divided into several sub-stages. Several types of skarns are distinguished: skarns where dominant mineral is hedenbergite (iron as $Fe^{2}+$, called "reduced skarns"), skarns with magnetite ($FeO \cdot Fe_2O_3$) and skarns with andraditic garnet (iron as Fe^{3+} called "oxidized skarns").

Post-skarn hydration stages: epidote replaces the garnet; amphibole (actinolite) replaces pyroxene; alteration of periclase into brucite, talc, serpentine on the olivine, chlorite, carbonate. Crystallization of hydrous minerals like pectolite. It is at this stage that the main mineralization (W, Sn, Be, Mo, B, Cu, Pb, Zn, Au, Ag, U, REE, etc.) associated with skarn, are formed.

The only known occurrence of charoite is rocks which have suffered a potassium metasomatism at the contact of a nepheline–aegyrine syenite of the Murun massiv, between the Chara and Olekma rivers in Yakutia, Russia.

Metamorphism and alterations of the ultrabasic rocks – rodingites

The moderate temperature metamorphism of ultramafic and ultrabasic (peridotite, serpentine) leads to the formation of rocks composed of talc, chlorite, amphibole (tremolite, anthophyllite), called steatite or soapstone.

Hydration and carbonation of serpentine form a paragenesis talc + magnesite. Brucite also appears in veins in serpentinite. Other carbonates (calcite, dolomite, and aragonite) may also appear.

Rodingites are rocks formed of grossular (or hydrogrossular), a pyroxene of the diopside–hedenbergite–aluminous diopside series and, in lesser quantities, vesuvianite, epidote, scapolite, ect, prehnite, magnetite/hematite. These rocks form veins and masses in massive serpentine. They likely originated from veins of gabbro or dolerite that underwent calcic metasomatism.

Very high pressure metamorphism

White schist facies, of very high pressure and low temperature is defined by the paragnenesis talc + kyanite:

$$\text{chlorite} + \text{quartz} \Leftrightarrow \text{talc} + \text{kyanite} + H_2O$$

3.4.2.3 Igneous rocks

Some of the minerals discussed in this chapter are found in various igneous rocks: allanite is common accessory mineral in intermediate to acidic igneous rocks rich enough in calcium (granite, granodiorite, quartz diorite, syenite, monzonite); scapolite exceptionally appears in some pegmatites.

Other calcic and magnesian minerals that occur in igneous roks, are specific to alkaline rocks, many of them strongly under-saturated in silica. Alkaline rocks are under-saturated in aluminum mainly in respect to alkalis (sodium and/or potassium), there are also rocks under-saturated in aluminum in respect to calcium.

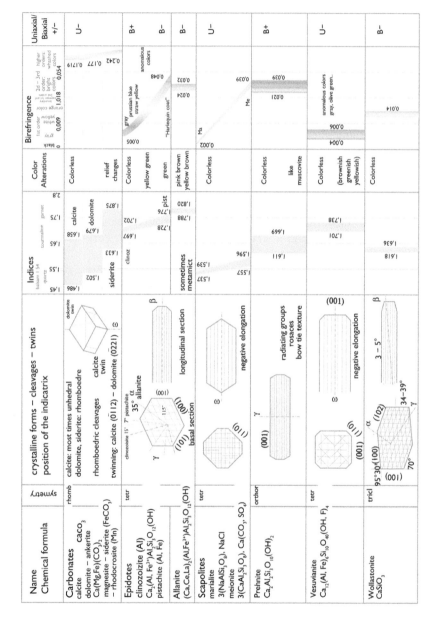

Name / Chemical formula	symmetry	crystalline forms – cleavages – twins / position of the indicatrix	Indices	Color Alterations	Birefringence	Uniaxial/Biaxial +/−
Carbonates CaCO₃ calcite; dolomite – ankerite Ca(Mg,Fe)(CO₃)₂; magnesite – siderite (FeCO₃); – rhodocrosite (Mn)	rhomb	calcite: most times unhedral; dolomite, siderite: rhomboedre; rhomboedric cleavages; twinning: calcite (01̄12) – dolomite (02̄21)	calcite 1.486 / 1.658; siderite 1.502 / 1.875; dolomite 1.679	Colorless; relief changes	0.009 ... 0.177 ... 0.242 ... 0.054	U−
Epidotes clinozoisite (Al) Ca₃(Al, Fe³⁺)Al₂Si₃O₁₂(OH); pistachite (Al, Fe)	tetr	clinozoisite 15° 7° pistachite 35° allanite; basal section (100)(001); β α γ 115°	clinoz 1.633; 1.697 / 1.702; pist 1.776	Colorless; yellow green	gray prussian blue straw yellow; anomalous colors	B+ B−
Allanite (Ca,Ce,La)₂(Al,Fe³⁺)Al₂Si₃O₁₂(OH)		longitudinal section	sometimes metamict 1.537 / 1.539; 1.728 / 1.788; 1.820	pink brown yellow brown; green	0.005 0.024 0.032	B−
Scapolites marialite 3(NaAlSi₃O₈)·NaCl; meionite 3(CaAl₂Si₂O₈)·Ca(CO₃, SO₄)	tetr	negative elongation ω γ (001)(01̄1)	1.557 Ma; 1.596 Me	Colorless	0.002 Ma ... 0.021 Me ... 0.039	U−
Prehnite Ca₂Al₂Si₃O₁₀(OH)₂	orthor	radiating groups rosaces bow tie texture (100) ω	1.611 / 1.669	Colorless; like muscovite	0.039	B+
Vesuvianite Ca₁₉(Al, Fe)₅Si₁₀O₄₀(OH, F)₄	tetr	negative elongation ω	1.701 / 1.738	Colorless (brownish greenish yellowish)	0.004 0.006 anomalous colors gray, olive green	U−
Wollastonite CaSiO₃	tricl	3 – 5° 34–39° 95°30′ (100) (1̄02) 70° (001) β γ α	1.618 / 1.636	Colorless	0.014	B−

Figure 3.37 Summary of the characters of: calcic and magnesian minerals.

An example of the latter group is given by some alkaline granites (for instance in Corsica), where the under-saturation in aluminum is marked by the presence of wollastonite instead of anorthite. Some phonolites and ijolites (igneous rocks with nepheline + augite) containing wollastonite are other examples.

Melilite, monticellite, titaniferous (and zirconian) garnet, titaniferous humites, perovskite ($CaTiO_3$, discussed with accessory minerals, §3.5.4) and carbonates are more or less important components of rocks strongly under-saturated in silica: alkaline volcanic and hypabyssal rocks, and their plutonic equivalents (when they exist). These alkaline rocks are classified into:

- Sodic rocks: nephelinites;
- Potassic rocks: leucitites, mafurites (rocks with kalsilite $KAlSiO_4$);
- Potassic rocks rich in biotite: kimberlites, alnöites (lamprophyres with melilite), lamproites;
- Calcic rocks: mélilitites (lavas containing more than 10% of melilite) and carbonatites.

Carbonatites are plutonic and volcanic rocks made mainly of carbonates: calcite (coarse-grained sövite and fine-grained alvikite), dolomite (beforsite), iron-bearing carbonates. Lavas made of sodium carbonate (natrocarbonatite) are known but, as this mineral is soluble in water, these rocks are not preserved.

Garnets of the andradite–melanite (Ti-andradite) series also occur in nepheline syenite and some phonolites. Vesuvianite, pectolite and calcite are also constituents of nepheline syenite. Pectolite also occurs in some rare mica peridotite.

Calcite, aragonite associated with zeolites, pectolite (associated with datolite, presented in §3.5.2.3) may crystallize in vesicules of basalts, andesites and their hypabyssal equivalents.

3.4.2.4 Hydrothermal veins

Carbonates (calcite, dolomite, ankerite, siderite, rhodocrosite) frequently appear as gangue minerals in ore veins (or massive orebodies). Some, such as siderite, may be mined.

Epidote, prehnite (with quartz, albite, carbonate, etc.) form of low temperature hydrothermal veins in crystalline terranes (a famous example are the alpine clefts).

The aragonite may be formed in the hot springs.

3.4.2.5 Alteration minerals

Saussurite is an alteration of calcic plagioclase into epidote (+ albite) (§3.1.2.3).

Scapolite occur as metasomatic replacement of plagioclase in basic igneous rocks (gabbros, dolerites, ophites the Pyrenees) and in some gneisses of high metamorphic grade.

Some partially altered biotites being shown in their cleavage lenses of epidote or prehnite.

3.5 ACCESSORY MINERALS

We call here accessory minerals, minerals bearers of elements relatively minor in the rocks: boron, phosphorus, titanium, zirconium, etc. Although these elements are often significant, they remain relatively minor to major elements such as Si, Al, Fe, Mg, Ca, Na, K. Some previously treated minerals fall within this definition: for example allanite (REE), piemontite (Mn) and beryl (Be).

We also deal here with spinels, which do not fit the above definition, because they could not conveniently be included in the preceding groups.

3.5.1 Spinel group

3.5.1.1 Chemical composition

Spinels s. l. are cubic minerals with the unit cell formula:

$$R^{2+}_8 \, R^{3+}_{16} \, O_{32}$$

Cations are distributed into eight tetrahedral A sites and 16 octahedral B sites. There are two structural types, the normal spinels which contain 8 R^{2+} in the A site and 16 R^{3+} in B site, and inverse spinels with 8 R^{3+} in A site and 8 R^{2+} + 8 R^{3+} in B site B.

More simply the formula may be written:

$$R^{2+} R^{3+}_2 O_4 \quad \text{ou} \quad R^{2+}O \, R^{3+}_2 O_3$$

with

R^{2+} = Fe^{2+}, Mg, Mn, Zn, Ni, Co, Ca
R^{3+} = Al, Fe^{3+}, Cr, V, Mn^{3+}

Titanium, a tetravalent element, is important in the spinel group; the Ti- end member has an inverse structure and its formula is $R^{2+}_2 Ti^{4+}O_4$ with mostly R^{2+} = Fe (*ulvospinelle*).

Most end members have been named, some are rare and of limited interest. There are solubility gaps between these end members. They are incompletely explored.

The spinel group minerals are classified according to R^{3+}.

$R^{3+} = Al$ – spinels group

There is a continuous solid solution between $MgAl_2O_4$ *spinel* and $FeAl_2O_4$ *hercynite*; the intermediate Mg-rich terms are called *pleonast*.

There is a continuous series between $MgAl_2O_4$ spinel and the zincian end member, *gahnite* $ZnAl_2O_4$ and between spinel and the manganiferous end member *galaxite* $MnAl_2O_4$. In this last series, ferric iron replaces aluminum. The latter two minerals are very rare.

Above 858°C, there is also a continuous series, between hercynite $FeAl_2O_4$ and magnetite $Fe^{2+}Fe^{3+}_2O_4$. At lower temperatures there is a solvus.

The intermediate terms $Fe^{2+}(Fe^{3+},Cr^{3+})_2O_4$ of the series between hercynite and chromite $Fe^{2+}Cr^{3+}_2O_4$ are called *picotites*.

$R^{3+} = Fe3+$ – magnetite group

Magnetite $Fe^{2+}Fe^{3+}_2O_4$ or Fe_3O_4 is a major constituent of the rocks. In magnetites, Ni, Co, Zn, Mg, Mn, Ca can substitute to Fe^{2+} ($MgFe^{3+}_2O_4$ is *magnesioferrite*) and Cr, V, Mn^{3+} to ferric iron.

There is a continuous series between magnetite and *ulvospinelle* $Fe^{2+}_2TiO_4$. Most magnetites of the igneous rocks contain titanium. Titaniferous magnetite and ilmenite $FeTiO_3$ (which forms a continuous series with hematite Fe_2O_3) are frequently associated, and the partition of titanium and ferrous/ferric iron between the two minerals is used to estimate temperature and oxygen fugacity.

Some natural magnetites contain an excess of ferric iron, indicating a solid solution between magnetite and *maghemite* γFe_2O_3, which is a low temperature polymorph of hematite. Maghemite is the produce of supergene alteration of magnetite. This mineral is unstable and transforms into hematite at increasing temperatures.

Other members of this group are rather unimportant: franklinite $ZnFe^{3+}_2O_4$ (with $Zn \Leftrightarrow Mn^{2+}$ substitution), jacobsite $Mn^{2+}Fe^{3+}_2O_4$ (with Mn^{2+} $\Leftrightarrow Fe^{2+}$, Mg and $Fe^{3+} \Leftrightarrow Mn^{3+}$ substitutions), trevorite $NiFe^{3+}_2O_4$ (with $Ni \Leftrightarrow$ Fe^{2+}, Mg substitution).

$R^{3+} = Cr$ – chromite group.

There is a continuous series between chromite $Fe^{2+}Cr_2O_4$ and magnesiochromite $MgCr_2O_4$. Chromium can also be replaced by aluminum and ferric iron. Zinc may substitute to ferrous iron in small quantities. The general formula of chromite is:

$$(Fe^{2+}, Mg, Zn)(Cr, Al, Fe^{3+})_2O_4$$

Compositions of chromites depend on the type of deposit. Aluminous chromites are used in the refractory industry (and to a lesser extend in metallurgy), chromites rich in chromium, in metallurgy.

Under the microscope, spinel is colorless. Spinels containing iron or titanium are colored in green, gray green, brown, green, brown, reddish brown to opaque. Magnetite and chromite are opaque and cannot be determined with the petrographic microscope.

3.5.1.2   Occurrences

Metamorphic rocks

Minerals of the spinel–hercynite group are minerals of high grade metamorphism. Spinel *s.s.* appears in calcareous-dolomitic rocks. The spinels of the spinel–pleonaste–hercynite group occur in pelitic rocks. Spinel is incompatible with quartz, spinel + quartz association being replaced by sapphirine or cordierite. Hercynite is compatible with quartz.

Magnetite characterizes a large family of skarns characterized by an oxygen fugacity intermediate between that of "reduced" skarns with hedenbergite (ferrous iron) and "oxidized" skarns with andradite (ferric iron). Rather pure, titatium-poor magnetite are mostly developed in exoskarns. Magnetites of endoskarn are characterized by high titanium content.

Hematite and limonite of the sediments and iron formations are reduced to magnetite during metamorphism. Magnetite is an essential constituent of banded iron formations in particular.

Igneous rocks

During fractional crystallization, chromites crystallize in basic and ultrabasic terms, magnetites appear in more evolved terms. In both cases these minerals evolve from magnesium-richer to iron-richer members.

Spinelles of kimberlites and lherzolites in enclaves in basalts belong to a magnesiochromite (Mg, Cr)–pleonastes (Fe, Cr)–picotite (Mg-Fe, Al) series.

Alpine peridotites are one of the two major types of chromium deposits (for example: New Caledonia, Cuba, Turkey, Albania) (Bouladon, 1986). Chromite form stratiform orebodies in peridotitic and/or gabbroic cumulates. The ore is either massive or made of crystals or nodules of chromite dispersed in a matrix of silicates ("leopard" or "antileopard" structures). It is an aluminous and magnesian chromite (refractory chromite) which may contain inclusions of heavy PGMs (Pt, Ir, Os). "Podiform" orebodies form irregular masses in harzburgites. Such bodies are often of tectonic origin (with perhaps a hydrothermal reworking). They are made of metallurgical chromite rich in iron, low in aluminum. Cu, Ni, Co are recovered as byproducts from the processing of these chromites (these elements are presumably contained in associated sulfides).

Mafic–ultramafic complexes of the Bushveld type are the other type of chromium deposit (Willemse, 1969, Von Gruenenwaldt et al., 1985). In the

Bushveld Igneous complex, chromite form layers (up to 1–2 m of thickness) associated with pyroxenites and anorthosites in the Critical Zone, 21 layers have been counted in the district of Lydenburg. That is a metallurgical chromite rich in iron, low in aluminum and magnesium, sometimes containing PGMs. The chromium contents ranges from 49 wt% Cr_2O_3 at the base of the critical zone to 43 wt% on top. The upper part of the Bushveld (Upper Zone: gabbros, ferro-gabbro) contains layers of magnetite mined for iron and vanadium. The contents of titanium and vanadium of this magnetite range from 12 wt% TiO_2, 2 wt% V_2O_5 at the base of the Upper Zone, to respectively 18 wt% and 0.3 wt% at the top.

Titaniferous magnetite is a ubiquitous mineral in igneous rocks. In the alkaline series of Velay (France), content in TiO_2 of the magnetites is 21–19 wt% in basalts/basanites, 23.5–23 wt% in hawaiites, 23–15 wt% in mugearites, 21–14 wt% in benmoreites, 20–15 wt% in trachyte and quartz trachyte, 14 wt% in phonolites. The content in Al_2O_3 decrease from 5.2–5.5 wt% in basalts/basanites to 2–0.5 wt% in the trachytes and quartz trachytes and 0.10 wt% in phonolites (unpublished data of E. Berger).

In plutonic rocks magnetite is easily re-equilibrated when the temperature drops, sometimes with exsolution of ilmenite and/or hematite. Ishihara (1977) distinguishes M magnetite (and ilmenite) granites and I granites with only ilmenite. These types are often identified to respectively I (igneous) granites and S (sedimentary) granites of Chappell and White (1974).

3.5.2 Boron minerals

3.5.2.1 *Tourmaline*

Tourmaline is a cyclosilicate formed of Si_6O_{18} rings alternating with layers of complex structure. These layers consist of three central octahedral Y sites surrounded by six Z sites (distorted octahedra), themselves connected by three boron ions. These layers are connected by tetrahedral X sites. OH can substitute for oxygen. The general formula of tourmaline is:

X Y_3 Z_6 B_3 Si_6 $(O, OH)_{30}$ (OH, F)
X = Na most times, sometimes, Ca
Y = Fe^{2+}, Mg, Li, Al, Li, Fe^{3+}
Z = Al and rarely Mg, Fe^{3+}, Mn, Cr

Most tourmalines belong to the following series:

- iron–tourmaline (schorl) Na $Fe^{2+}_3Al_6$ – magnesian–tourmaline (dravite) $NaMg_3$ Al_6;
- iron–tourmaline – lithium–tourmaline (elbaite) $Na(Li, Al)_3$ Al_6

Most common tourmalines are iron tourmalines.

Tourmaline is a clastic mineral reworked in sediments. It is common in metapelites without necessarily coming from former clastic tourmalines: indeed, boron contained in sea water is adsorbed by clays and recrystallization, even under diagenetic or anchimetamorphism conditions, induces the formation of tourmaline (which can also nucleate on former clastic tourmalines).

Boron is an hygromagmaphile element which is concentrated in the evolved magmas and pneumatolytic and hydrothermal fluids. Tourmaline is a mineral that crystallizes in evolved granites, pegmatites and aplites. Most often it is an iron-tourmaline. Some evolved granites and pegmatite contain lithium-tourmaline. A single crystal can also show a zonation between iron-tourmaline and lithium-tourmaline.

Percolation of boron-bearing hydrothermal fluids produces crystallization of tourmaline in the host rocks that are more or less completely replaced, the textures being preserved. The result may be massive tourmalinites. This boron metasomatism occurs particularly around some ore veins, for example in tungsten and tin deposits.

Magnesian tourmaline occurs mainly in metamorphic rocks derived from more or less impure calcareous-dolomitic sediments. It is probable that these magnesian tourmaline are more commonly calcic (uvite, $(Ca, Na) Mg_3Al_6)$ than sodic (dravite, $NaMg_3Al_6$).

3.5.2.2 Axinite

Axinite is a rare sorosilicate made of almost plane layers composed of double Si_2O_7 tetrahedra linked by two tetrahedra containing boron, alternating with layers made of chains of 6 octahedra containing aluminum linked by polyhedral sites containing CaO_6 and $CaO_5(OH)$; calcium can be replaced by Fe, Mg, Mn. Hence the formula:

$$H(Ca, Fe, Mg, Mn)_3Al_2BSi_4O_{16}$$

The axinite appears in calcareous (or mafic) rocks which underwent boron metasomatism (contact metamorphism and skarns). It is also a mineral of the Alpine clefts with epidote, albite, etc.

3.5.2.3 Datolite

Datolite $CaSiO_4B$ (monoclinic) is a nesosilicate which mostly appears as a hydrothermal (and/or secondary) mineral in the vesicles and in veins in basic rocks (and in skarns and serpentinites) with calcite, prehnite, axinite, zeolites, etc.

3.5.3   Phosphates

3.5.3.1   *Apatite*

Apatite $Ca_5(PO_4)_3(OH, F, Cl)$ is an accessory mineral common in all types of rocks. Apatite varies mostly by substitution:

$$F \Leftrightarrow OH \Leftrightarrow Cl$$

The end members are called fluor-apatite/hydroxyl-apatite/chlor-apatite. Fluor-apatite and fluor-hydroxylapatite are the most common.

There is also a substitution:

$$PO_4^{3-} \Leftrightarrow CO_3^{2-} + F^-$$

Francolite is an apatite rich in CO_3^{2-} and fluorine.

Apatite may contain silica in substitution to phosphorus; the charge deficit is compensated either by Rare Earth (or thorium) as a substitution to calcium:

$$Ca^{2+} + PO_4^{3-} \Leftrightarrow REE^{3+} + SiO_4^{4-}$$

or by the coupled substitution of phosphorus by sulfur which forms a continuous solid solution toward ellestadite ($Ca_5(SiO_4, SO_4)_3(OH, Cl, F)$):

$$2 PO_4^{3-} \Leftrightarrow SO_4^{2-} + SiO_4^{4-}$$

Apatite is an ubiquitous mineral.

Fluor-apatite (hydroxy-apatite and fluor-) is found in both basic and acidic igneous rocks, including pegmatites. It appears, however, only beyond some grade of differentiation. Thus, in the Bushveld (tholeiitic series) it begins to crystallize in the upper part of the Upper Zone. In the alkaline volcanic series of Velay, it is missing in basalts/basanites and poorly differentiated hawaiites, and it appears in differentiated hawaiites and is subsequently present in mugearites, benmoreites, trachytes and phonolites.

Apatite is very abundant in differentiated alkaline rocks, particularly in plutonic rocks. It is a major constituent of some rocks of carbonatite complexes: at Palaborwa (/Phalaborwa, Transvaal): it is mined in apatite–magnetite–phlogopite–calcite–apatite rocks and olivine–phlogopite–clinopyroxene–olivine rocks. Alkaline rocks and carbonatite complexes provide 23% of world production of phosphates.

Fluor-apatite is the major constituent of bones and teeth of vertebrates and fish scales.

The major deposits of sedimentary phosphates (76% of world production) are sediments of areas of upwelling in the coastal shelves. Fragments of bones and fish scales serve as nuclei for precipitation of apatite (rich in CO_3^{2-}). Cryptocrystalline phosphates are called collophane. They are mainly made of apatite, but their exact composition is uncertain. Apatite also forms nodules and crusts in the anoxic zone on the continental slopes.

Phosphorites (and guano) are the accumulation of debris and excrement of vertebrates trapped in karsts; their mining interest is no more than historical (1% of world production).

Apatite is a common accessory in metamorphic rocks as well as of ortho- and para-origin. It also occurs in hydrothermal alpine clefts.

3.5.3.2 Monazite

Monazite (Ce, La, Th) (PO_4) is the bearer of REE in calcium-poor granites, syenites and pegmatites. Some monazites are rich in europium. Monazites also contain uranium, aluminum and ferric iron in very small quantities. Monazite is used for radiometric U/Pb dating. The obtained ages are generally concordant.

3.5.3.3 Xenotime

Xenotime YPO_4 (tetragonal) is a rare phosphate of Rare Earths (mostly heavy lanthanides: dysprosium, erbium, terbium and ytterbium) that also contain uranium and thorium. It may also contain traces of calcium, silicium and arsenic. It is mostly a mineral of granite pegmatites and evolved igneous rocks. It may be a clastic and may be concentrated in black sands with other heavy minerals. Diagenetic xenotime is also reported, particularly as coating on detrital zircons. It is an ore of heavy Rare Earths.

3.5.4   Lithium bearing minerals

These minerals appear mainly in lithian pegmatites; they are also found in associated evolved granites, greisens and tin veins. Some are presented in other parts of this book. These are: lithium micas (§3.2.1.5), spodumene (§3.2.4.1), amblygonite, petalite, elbaite (lithian tourmaline §3.5.2.1). Pollucite, cassiterite (§3.7), the potassium feldspars, orthoclase and microcline, albite and quartz may be associated to these lithium minerals.

3.5.4.1 Amblygonite

Amblygonite (Li, Na) $AlPO_4$ (F, OH) is a fluophosphate which forms a series with the montebrasite, term low in fluoride. These triclinic minerals can be confused with feldspars but are distinguished under the microscope by their cleavages and by their alterations. These are lithium ore and some crystals may be used as gemstones.

3.5.4.2 Petalite

Petalite $LiAlSi_4O_{10}$ is a phyllosilicate which crystallizes in the monoclinic system as tabular crystals flattened on (001). Na or K may be substituted to Li and iron in very small quantities to the Al. Petalite is altered into montmorillonite.

3.5.4.3 Pollucite

Pollucite $(Cs, Na)AlSi_2O_6 \cdot H_2O$, contains no lithium, but it is associated with lithium-bearing minerals in the same occurences. The mineral is cubic, is most times enhedral and has no cleavage; its index (1.520) is significantly lower than the one of quartz. Pollucite forms a series with analcime ($NaAlSi_2O_6 \cdot H_2O$) (§3.1.3) but the occurrences of these minerals are quite different.

3.5.5 Titanium bearing minerals

Biotite is probably, with ilmenite, the major bearer of titanium in many rocks. Recall that biotite may contain up to 5–6 wt% TiO_2. Titaniferous magnetite is also an important bearer of titanium in igneous rocks.

Titanium has also its own minerals:

- **rutile** TiO_2 (tetragonal, positive unixial) and its low temperature polymorphs; **anatase** (tetragonal, negative uniaxial) and **brookite** (orrthorhombic);
- **ilmenite** $FeTiO_3$;
- pseudobrookite a continuous series between $Fe_2O_3 \cdot TiO_2$ end member and $FeO \cdot 2TiO_2$ end member; very rare tardimagmatic mineral of the volcanic alkaline rocks;
- **titanite** $CaTiSiO_4(O, OH, F)$;
- **perovskite** $CaTiO_3$.

3.5.5.1 Rutile

TiO_2 may contain Fe^{3+}, Fe^{2+}, Al, Cr, V, Nb, Ta, Sn in minor amounts or trace amounts. It is the form of high temperature of the polymorphs of TiO_2 as it has the smallest molar volume of the three polymorphs, it is also the form of high pressure. It thus occurs in eclogites and of high pressure amphibolites.

Rutile is less common than ilmenite in igneous rocks, but occurs in some granitic pegmatites and anorthosites. It is rare in volcanic rocks.

Chlorite derived from titaniferous biotites commonly contains rutile needles, sometimes grouped into 6 pointed stars, as sagenite twins.

Rutile appears as inclusions in other minerals such as quartz, as fine needles sometimes grouped into 6 pointed stars. It is thus responsible for the effect of asterism of star sapphires.

Forms of low-temperature anatase and brookite are very often confused with rutile under the microscope. These minerals occur in igneous and metamorphic rocks. They are mostly minerals of alteration of other titaniferous minerals, titanite and ilmenite.

3.5.5.2 Ilmenite

Ilmenite $FeTiO_3$ may contain limited amounts of magnesium and manganese in substitution to iron. There is a complete solid solution between ilmenite $FeTiO_3$ and hematite Fe_2O_3 at temperatures above 1050°C. At lower temperatures, there is a solvus and ilmenite commonly contain up to 5% Fe_2O_3. The common paragenesis ilmenite (containing hematite in solid solution) – titaniferous magnetite allows a determination of the temperature and oxygen fugacity. Ilmenite may also contain traces of V, Cr, Al.

The mineral is opaque, and cannot be precisely determined with the petrographic microscope.

Ilmenite is a ubiquitous mineral in igneous rocks. Ilmenites from kimberlites and peridotite xenoliths may contain up to 12–13 wt% MgO · So Mg-ilmenite is a guide for diamond exploration. In anorthosite, ilmenite associated with rutile commonly form cumulative bodies likely to be of economic interest (for example, Quebec, New York state, Mexico). In the alkaline series of Velay, ilmenite only appears in differentiated basalts/basanites, hawaiites and mugearites. Its molar contents of hematite are growing from 6 to 8% from the basalts to the mugearites. It is absent in more evolved terms. Manganese-bearing ilmenites are found in differentiated igneous rocks and carbonatites.

Ilmenite is a mineral easily reworked in the sediments. With rutile it belongs to the family of heavy minerals concentrated in beach sands, which may sometimes have an economic interest (Florida, India, Senegal, Madagascar).

Ilmenite is a common mineral in metamorphic rocks as well as of ortho- and para-derivation. At high pressures (eclogite facies and granulite facies of high pressure), ilmenite is replaced by rutile according to reactions such as:

$FeTiO_3 + SiO_2 \Leftrightarrow FeSiO_3 + TiO_2$
ilmenite + quartz = orthopyroxene + rutile (Hayob et al. 1993)
$3FeTiO_3 + Al_2SiO_5 + 2SiO_2 \Leftrightarrow Fe_3Al_2Si_3O_{12} + 3TiO_2$
ilmenite + sillimanite + quartz $\Leftrightarrow$ almandin + rutile (Feenstra and Engi, 1998)
$CaFeSi_2O_6 + 2CaAl_2Si_2O_8 + 2FeTiO_3 \Leftrightarrow$
$Ca_3Al_2Si_3O_{12} + Fe_3Al_2Si_3O_{12} + 2TiO_2$
clinopyroxene + plagioclase + ilmenite $\Leftrightarrow$ Ca-Fe garnet + rutile
(Mukhopadhyay et al., 1992)
$3CaFeSi_2O_6 + 3CaAl_2Si_2O_8 + 3FeTiO_3 \Leftrightarrow Ca_3Al_2Si_3O_{12} + 2Fe_3Al_2Si_3O_{12}$
$+3CaTiSiO_5$
clinopyroxene + plagioclase + ilmenite $\Leftrightarrow$ Ca-Fe garnet + titanite
(Mukhopadhyay et al., 1992)

3.5.5.3 *Titanite*

Titanite $CaTiSiO_4(O, OH, F)$ may contain aluminium and ferric iron in substitution to titanium:

$$Ti^{4+} + O^{2-} \Leftrightarrow (Al, Fe)^{3+} + (F, OH)^-$$

Rare Earths, ferrous iron, magnesium may also replace calcium.

There may be traces of niobium, tantalum, vanadium, uranium and thorium. The presence of these radioactive elements allows the use of titanite for radiometric dating. Their radiation can also destroy the lattice of titanite (metamict state).

Titanite is a common mineral in the acidic to intermediate igneous rocks, provided they contain sufficient calcium. It is relatively abundant in nepheline syenite. Titanite is also common in metamorphic rocks rich enough calcium, especially some metapelites and particularly amphibolites and calc-silicate-gneisses.

3.5.5.4 *Perovskite*

Perovskite is orthorhombic pseudo-cubic mineral, which has a very compact structure: it is the type of stable minerals in mantle conditions. Its simplified formula is $CaTiO_3$. Calcium is commonly replaced by Rare Earths (dominant cerium and less abundant lanthanum) and yttrium (minor), and to a lesser extent by Fe^{2+}, Sr and/or Na. Titanium is replaced by niobium, often associated with minor amounts of zirconium, and to a lesser extent, tantalum:

$$(\mathbf{Ca}, Ce, La, Y, Fe^{2+}, Sr, Na) (\mathbf{Ti}, Nb, Ta, Zr) O_3$$

Perovskite is under-saturated in silica (quartz + perovskite = titanite).

It is a mineral of alkaline igneous rocks under-saturated in silica, where it is associated with nepheline, leucite, melilite. It is often found in mafic and ultramafic: rocks like ijolites, jacupirangites, nephelinite, kimberlites and in the carbonatites. It is also found in nepheline syenites and their pegmatites. It is known in meteorites.

It is also found in skarns.

3.5.6 Zircon

Zircon $ZrSiO_4$ is a common accessory mineral of the igneous rocks. It is more common in intermediate to acidic rocks. It is relatively abundant in differentiated alkaline rocks such as nepheline syenite.

Its crystals may be independent (evolved granites, pegmatites, nepheline syenites) or be inclusions in other minerals (biotite, amphibole, cordierite). In this case, pleochroic halos due to radiation from radioactive trace elements in zircon develop in the hosts: the dark halo in biotite and amphibole, yellow in cordierite.

Pupin and Turco (1972) have shown some correlation between the form of single crystals and type of magmatism. These results are far, however, of having an absolute and systematic value.

Zircon is one of the heavy minerals easily reworked – sometimes concentrated – in the sediments. It is therefore commonly found in metamorphic rocks as well as ortho- as para-origin.

Zircons commonly contain 1 wt% of hafnium, and traces of heavy Rare Earths, yttrium, uranium and thorium. Zircon is commonly used in geochronology due to the presence of these radioactive elements.

Generally a single crystal of zircon records a complex history with several episodes of growth (formation of zoned euhedral crystals) and erosion (eroded grain boundaries intersecting the previous zonation). Zircon growth occurs either when the zircon is incorporated into a newly formed igneous rock, or during metamorphism. The core of the crystals is not re-equilibrated. During the alteration of these rocks, the zircon crystals may be eroded, re-sedimented and subsequently included in a new cycle (inherited zircons): *zircons are forever*!

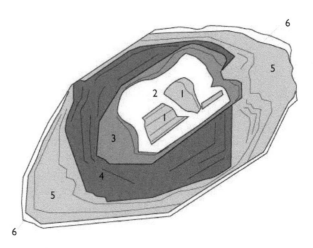

Figure 3.38 Image by catodoluminescence of a zircon from the orthogneiss of Canigou (Pyrénées orientales, France; after Cocherie, 2005). The size of this crystal is of the order of 250 microns. Zone 1 is the oldest part of the zircon. Then comes a zone of growth (2). This first zircon is eroded, then comes a second zone of growth (3). Then we notices the stages of erosion followed by growth (4) and (5) and an area of subsequent marginal overgrowth (6).

So one can understand that the ages obtained on a population of zircons or even on a single crystal of zircon are discordant or of uncertain interpretation. The current method used involves single point separate measurements in each zone recognized in a single crystal age. For example, this method provided on reworked zircons in metapelites of the garnet zone of Montagne Noire following ages (Ma) (Gebauer et al., 1989): 3148, 2900–2765 (Archean), 2140–1750 (Paleoproterozoic), 1000 (Mesoproterozoic), 600–556 (Neoproterozoic, Cadomian orogeny), 435 (Silurian, Caledonian orogeny). Debate on the significance of these ages remains open.

3.5.7   Titano- and zircono-silicates and silicates of the alkaline rocks

In (plutonic and/or volcanic) peralkaline rocks, the excess of alkali over aluminum causes the appearance of alkali ferro-magnesian minerals (sodic amphiboles and pyroxenes), and among the accessory minerals, of sodic titano- silicates and zircono-silicates.

Here are some more or less common minerals of this group (the ones shown on the CD are indicated with an asterix*):

- aenigmatite* $Na_2Fe^{2+}{}_5Ti\,Si_6O_{20}$ (triclinic)
- astrophyllite* $(K, Na)_3(Fe^{2+}, Mn)_7Ti_2\,(Si_4O_{12})_2$
  $(O, OH, F)_7$ (triclinic)
- lamprophyllite $Na_2(Sr, Ba)_2Ti_3(SiO_4)_4(OH, F)_2$ (monoclinic)
- rinkite $Na(Na,Ca)_2(Ca, Ce)_4Ti(F_2/(O, F)_2/(Si_2O_7)_2)$ (monoclinic)
- mosandrite $(Na,Ca,Ce)_3Ti\,(SiO_4)_2\,F$ (monoclinic)
- eudyalite* $(Na, Ca, Fe^{2+})_6\,Zr\,(Si_3O_9)_2(OH, F, Cl)$ (trigonal)
- catapleiite $(Na,Ca)_2\,Zr\,(Si_3O_7)\,SiO_4\cdot 2H_2O$ (monoclinic)
- låvenite* $(Na, Ca, Mn, Fe)_3(Zr, Nb, Ti)$
  $Si_2O_7\,(OH, F)$ (monoclinic)
- wöhlerite* $Ca_2\,Na\,(Zr, Nb)\,Si_2O_7\,(O, OH, F)_2$ (monoclinic)
- rosenbuchite $(Na, Ca, Mn)_3(Zr, Ti, Fe^{3+})$
  $(SiO_4)_2\,(OH, F)$ (triclinic)

There are others ...

These are minerals of nepheline syenites and associated pegmatites. Most of them occur in rocks under-saturated in silica. Aenigmatite, astrophyllite (and very rarely, eudyalite, låvenite and catapleiite) are also found in peralkaline granites. Aenigmatite also occurs in volcanic rocks: alkaline rhyolite, trachyte, phonolite. Wöhlerite also occurs in fenites and carbonatites.

Name / Chemical formula	symetry	crystalline forms – cleavages – twins / position of the indicatrix	Indices (1,45 – 1,55 – 1,65 – 1,75 – 2,8; balsam 1,54, quartz, tourmaline, garnet)	Color / Alterations	Birefringence (black, gray, white, yellow, orange color, boundary between 1st and 2nd orders; 0,009 – 0,018 – 0,054)	Uniaxial/ Biaxial +/–
Tourmaline Na(Fe, Mg, Mn, Li)$_3$ $Al_6Si_6O_{18}(OH)_4(BO_3)_3$	Trig	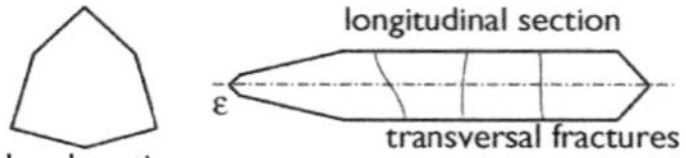 basal section; longitudinal section; ε; transversal fractures; negative elongation	1,639 – 1,657	green (blue, colorless zoned inverse pleochroisme	0,019 – 0,035	U–
Apatite $Ca_5(PO_4)_3(F, Cl, OH)$	hex	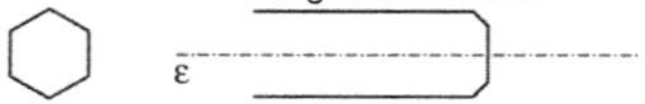 basal section; longitudinal section; ε; negative elongation	1,637 – 1,667	Colorless	0,002 – 0,005	U–
Zircon $ZrSio_4$	tetr	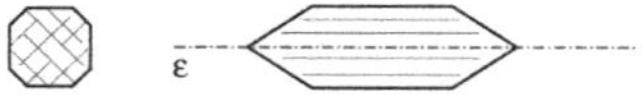 basal section; ε; positive elongation	1,926 – 1,985	Colorless (pleochroic halos)	0,039 – 0,059	U+
Titanite $CaTiSiO_4(O,OH,F)$	mnocl	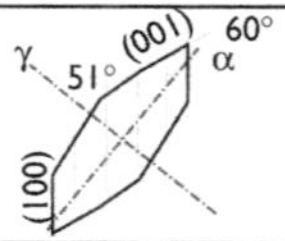 γ; 51°; (001); 60°; α; (100)	1,885 – 2,081	brownish (greenish, pink brown yellow)	0,108 – 0,160	B+
Rutile TiO_2	tetr	grains acicular crystals 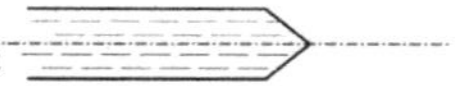 ε positive elongation sagenite	2,609 – 2,903	red brown yellow brown green brown	0,286–0,287	U+
Spinel $R^{2+}O{\bullet}R^{3+}{}_2O_3$ R^{2+} = Fe^{2+} Mg, Zn, Mn R^{3+} = Al, Fe^{3+}, Cr	cub	cubes – octaedra octaedral cleavage	1,718 – 2,05	Colorless blue green green brownish red brown opaque		

Figure 3.39 Summary of the characters of the accessory minerals.

3.5.8 Oxides of niobium, tantale and zirconium of alkaline rocks

- perovskite $CaTiO_3$ (cubic)
- baddeleyite ZrO_2 (monoclinic)
- pyrochlore* (Nb) – microlite (Ta) series
  $(Ca, Na)_2(Nb, Ta)_2O_6 (OH, F)$ (cubic)
- zirconolite* $(Ca, REE, U, Th)Zr$
  $(Ti, Nb, Ta)_2O_7$ (several polymorphs)

Perovskite is a mineral of rocks undersaturated in silica; it is discussed in §3.5.5.4.

Baddeleyite ZrO_2 (monoclinic) is a rare mineral of silica undersaturated rocks: syenites, gabbro, anorthosites, carbonatites, kimberlites. Baddeleyite is also a heavy mineral that are concentrated in some clastic sediments like black sands.

Pyrochlore and *microlite* $(Ca, Na)_2(Nb,Ta)_2O_6 (OH, F)$ (respectively Nb and Ta end members) are cubic minerals. They are found in alkaline silica-saturated or under-saturated pegmatites and in carbonatites. They are niobium – tantalum ores (Brazil, Quebec, Nigeria).

There are several polymorphs of formula $CaZrTi_2O_7$: the most common form is *zirconolite ss* (metamict); 3T zirconolite is trigonal, 3O zirconolite is orthorhombic, 2M zirconolite is monoclinic and zirkelite is cubic (Bayliss et al., 1989). Zirconolite *s.l.* may by confused under the microscope with rutile or titanite and its determination requires either the SEM or the electron microprobe. The precise determination of the different polymorphs requires to know the structure. This mineral is considered very rare. But it may be relatively common in alkaline rocks. Zirconolite is used to fix the actinides.

3.6 MINERALS OF SEDIMENTARY ROCKS AND ALTERITES

Clastic may contain any mineral reworked as clasts. Some groups of minerals are however more specific of sedimentary rocks: these are carbonates (discussed above), clays, iron and aluminum hydroxides (diaspore, α-AlO(OH), discussed above, boehmite β-AlO(OH), gibbsite $Al(OH)_3$), goethite α-FeO(OH), limonite $FeO(OH)$-nH_2O; non treated here) and evaporite minerals.

3.6.1 Clay minerals

3.6.1.1 Structure and chemical composition

The clays are hydrated aluminum silicates. They are sheet silicates (phyllosilicates) made of silicates sheets connected by interlayers.

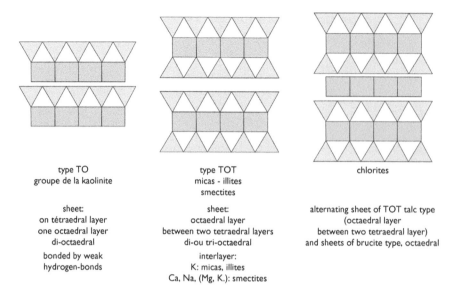

type TO	type TOT	chlorites
groupe de la kaolinite	micas - illites	
	smectites	

sheet:	sheet:	alternating sheet of TOT talc type
on tétraedral layer	octaedral layer	(octaedral layer
one octaedral layer	between two tetraedral layers	between two tetraedral layer)
di-octaedral	di-ou tri-octaedral	and sheets of brucite type, octaedral
bonded by weak	interlayer:	
hydrogen-bonds	K: micas, illites	
	Ca, Na, (Mg, K.): smectites	

Figure 3.40 Structure clay minerals and chlorites.

Two types of sheets occur:

1   TO type structures (also said 1/1), consisting of a tetrahedral layer and an octahedral layer; TO structure is characteristic of minerals of the kaolinite group.
2   TOT type structures (also said 2/1): two tetrahedral layers with inverted apexes, sandwiching an octahedral layer. The filling of the octahedral layer can be of 2 (dioctahedral sheets) or 3 (trioctahedral sheets) cations per unit cell. This type of structure occurs in micas, illites, smectites, vermiculites and chlorites.

The interlayer that connects the sheets is also of several types:

1   There is no interlayer; bonding between the sheets are hydrogen bonding; this occurs in kaolinite/dickite/nacrite, and, outside of the clay minerals, in talc and pyrophyllite.
2   Interlayer consists of potassium ions in micas and clays of the illite group; in the latter the interlayer is not complete.
3   In the group of smectites, interlayer has a very incomplete filling of Ca^{++} and/or Na^+ (and minor quantities of K, Cs, Sr, Mg, H), these ions are easily exchanged. These absorbing properties are used in industrial use of these minerals, such as in fuller's earth formed mostly of montmorillonites.

4   In vermiculite, the interlayer (filling also very incomplete) contains mostly exchangeable Mg^{2+} ions, and to a lesser extend of Ca^{2+} (and Na^+).

5   Halloysite (mineral of the kaolinite group), smectites and vermiculite contain incomplete interlayers of water. There is one layer in halloysite, two (or three) layers in smectite in the case of Ca ions, one in the case of Na ions, and two layers in vermiculite. This water is very weakly bound to the lattice and is easily lost when temperature rises (between 100–250°C for the major part in the smectites, 300°C in vermiculite). This loss of water causes a contraction of the network and exfoliation of vermiculite. Inversely, water can be easily adsorbed by these clays. This causes considerable variations of the interlayer distance (10–17.5 Å), hence the name of expanding-lattice clays.

6   In chlorite (which are sometimes classified with clay minerals) the sheets are separated by brucite-type layers $Mg(OH)_2$.

There are about 50 clay minerals. We merely indicate the major groups that are classified by their structure and chemical composition (Table 3.3):

Table 3.3 Clay minerals – comparison with other phyllosilicates and potassium feldspar.

	Interlayer cation	Interlayer water **	Octahedral sites	Tetrahedral sites	
*potassium feldspar**	K*			$AlSi_3$	O_8
muscovite	K		Al_2	$AlSi_3$	O_{10} $(OH)_2$
illite	K_{1-x}		Al_2	$Al_{1-x}Si_{3+x}$	O_{10} $(OH)_2$
kaolinite			Al_2	Si_2	O_5 $(OH)_4$
pyrophyllite			Al_2	Si_4	O_{10} $(OH)_2$
talc			Mg_3	Si_4	O_{10} $(OH)_2$
phlogopite	K		Mg_3	$AlSi_3$	O_{10} $(OH)_2$
smectites dioctahedral trioctahedral	$(1/2\ Ca, Na)_{0.35}$	nH_2O (two layers if Ca, one layer if Na)	$(Al, Mg, Fe)_2$ $(Mg, Fe, Al)3$	$(Si, Al)_4$	O_{10} $(OH)_2$
vermiculite	$(\underline{Mg}, Ca)_{0.3-0.45}$	nH_2O (two layers)	$(\underline{Mg}, Fe^{3+}, Al)_3$	$(Si, Al)_4$	O_{10} $(OH)_2$
palygorskite		$4H_2O$	$(Al, Mg)_2$	Si_4	O_{10} OH
sépiolite		$6H_2O$	Mg_4	Si_6	O_{15} $(OH)_2$
chlorites	brucite-type layer $(Mg, Fe, Al)_3(OH)_6$		$(Mg, Fe, Al)_3$	$(Si, Al)_4$	O_{10} $(OH)_2$

*The names of clay minerals are in regular characters; the names of other minerals are in italic.
* Potassium feldspar is a tectosilicate and the concept of interlayer has no sense. Its chemical formula is decomposed so it can be compared to other minerals in this table.
**Water in the ribbons of the fibrous clay minerals palygorskite and sepiolite.

- *kaolinite: group*, structure of TO (1/1) type; the only cations are Si and Al: kaolinite and its polymorphs (differing by the type of stacking of the sheets), dickite and nacrite have no interlayer; halloysite contains interlayer water;
- *illites group*, close to the micas of which they differ by a greater richness in silica and lower potassium content (substitution K Al^{IV} ⇔ Si □). Sericite is a fine-grained muscovite/illite, product of alteration of plagioclase. Glauconite is an illite containing ferric iron (ferrous iron and magnesium) in substitution to octahedral aluminum;
- *smectites* form a large family of TOT (2/1) type structure which contains Ca and Na in interlayer. Smectites are di-or trioctahedral. Among the dioctahedral smectites, beidellites are purely aluminous, montmorillonites contain magnesium in the octahedral site (and octahedral and tetrahedral aluminum), nontronite contains ferric iron in octahedral site (and aluminum in tetrahedral site). The principal trioctahedral smectite is saponite which contains magnesium in octahedral site;
- *vermiculites* are trioctahedral minerals of TOT (2/1) type structure; the interlayer is occupied by magnesium and to a lesser extent by calcium. The octahedral site is occupied by magnesium, with aluminum and ferric iron.
- *mixed layer clays* (in French: "interstatifiés" = interbedded) made of alternating sheets of more common minerals. Ordering of the sheets may be regular or random. The most common are illite–smectite and chorite–smectite;
- fibrous clays of the *palygorskite group* are made of a stacking of ribbons of TOT (2/1) type alternating with empty channels where zeolitic water may enter: Palygorskite (or attapulgite) is aluminous, sepiolite is magnesian.

In common argillaceous rocks in the current size of kaolinite crystals is about 5 microns, the illite crystals ranges from 0.1–0.3 microns, the montmorillonite cystals even smaller: **clay minerals cannot be determined under the petrographic microscope** but require the use of X-ray diffraction, thermal analysis, electron microscopy, etc. However we include in the CD a few clay minerals: glauconite recognizable by its color, kaolinite, which in the pores of sandstones form characteristic accordions-shaped crystal of relatively large size, etc.

3.6.1.2   Occurrences

Clays are *alteration* minerals: *hydrothermal* alteration, and on a much larger scale, *pedogenic* alteration.

The rocks of the Earth's crust are basic (silicates, carbonates) and reducing (ferrous iron, organic matter) media. Rainwater is oxidizing (presence of dissolved O_2) and acid (dissociation of dissolved CO_2: $CO_2 + H_2O$ ⇨ $H^+ + HCO_3^-$). Alteration results from the desequilibium between these two media.

There are hydrolysis reactions: oxygens of the lattice of the silicates preferentially bind to H^+ and cations are released. If one compares the compositions of feldspar, muscovite, illite and kaolinite, there is in this sequence a progressive leaching of potassium and generally of alkalis:

$$3KAlSi_3O_8 + H^+ + 12H_2O \Rightarrow KAl_2AlSi_3O_{10}(OH)_2 + 6Si(OH)_4 + 3K^+$$
$$\text{potassium feldspar} \quad \Rightarrow \quad \text{illite}$$

$$3KAlSi_3O_8 + 2H^+ + 9H_2O \Rightarrow Al_2Si_2O_5(OH)_4 + 4Si(OH)_4 + 2K^+$$
$$\text{potassium feldspar} \quad \Rightarrow \quad \text{kaolinite}$$

Such reactions may be written in a different way to highlight the role of dissolved CO_2:

$$CaAl_2Si_2O_8 + 2CO_2 \text{ (aq.)} + 3H_2O \Rightarrow Ca^{2+} + 2HCO_3^- + Al_2Si_2O_5(OH)_4$$
$$\text{Anorthite} \quad \Rightarrow \quad \text{kaolinite}$$

So feldspars are altered by leaching of the alkalis into illite group sheet minerals and then by a further leaching of the alkalis of interlayer into kaolinite. Similarly biotite is transformed into chlorite. The replacement of K^+ ions in interlayer site by Mg^{2+} and/or Fe^{2+} leads to vermiculites.

At a more advanced stage, the interlayer ions are replaced by ions of the sheet, and ultimately they are even evacuated. It is thus formed of mixed layer chlorite – smectite/vermiculite or illite – smectite/vermiculite. An extensive degradation leads to the formation of smectites and even a hydrous aluminum silicate gel or allophane.

The intensity of this alteration depends, of course, on the climate: temperature and precipitations.

Smectites derive from basic rocks, especially volcanic ash and tuff. Bentonites are rocks formed of montmorillonite and beidellite. Vermiculite comes from biotite/phlogopite in ultramafic or mafic rocks. Vermiculite from phlogopitites is exploited in the Palabora carbonatite complex. Plastic and expansion (exfoliation) properties associated with the hydrated interlayer of smectites and vermiculite are used in industry (drilling muds, ceramic, cement, etc.). This interlayer water may be replaced by alcohols and other organic compounds in industrial uses of these minerals.

The clays formed by weathering may be eroded and resedimented without transformation: many more or less pure clay formations are of purely *clastic* origin. When the eroded clay minerals are very degraded (very extensive leaching of the interlayer thorough, open sheets, etc.), these sedimentary minerals may be modified and transformed into more crystalline materials using ions encountered in the sedimentary basin, especially Mg^{2+} and/or K^+: this is the phenomenon of *aggradation recrystallization*.

Neoformation may also occur using ions (especially Mg) or colloids contained in the waters of the sedimentary basin, especially if it is subjected to high evaporation which concentrates the ions and colloids. Palygorskite and sepiolite are neoformed clays in conditions of evaporitic environment.

Kaolinite, illite and more rarely mixed layer clays, may be authigenous or formed during early diagenesis in sandstone pores. Kaolinite in particular has an accordion-shaped habit due to the stacking of flattened pseudohexagonal crystals.

Glauconite forms in aerobic littoral environments, perhaps in association with organic matter (foraminifera, pellets).

Diagenesis due to burial is well documented by the study of oil drilling. Diagenesis and anchimetamorphism produce:

- a departure of the interlayer water (which may have a role in the migration of oil and gas);
- a gradual recrystallization of illite;
- mineralogical transformations: smectites become mixed layer clays and then illite and chlorite, kaolinite is replaced by dickite and nacrite, and later by chlorite and illite. Smectites begin to disappear at temperatures of 70–95°C, that is, in the case of a normal geothermal gradient, at a depth of 2–3 km. These changes suppose a supply of K^+ ions. The origin of which is generally attributed to the destruction of clastic feldspars; but, as many argilites or shales do not contain this mineral, perhaps the origin of this potassium should be in connate waters.

Finally the only minerals that are stable under the conditions of anchimetamoprhism at the beginning of metamorphism are illite and chlorite.

3.6.2   Evaporites minerals

The principal minerals of this group are gypsum $CaSO_4 \cdot 2H_2O$, anhydrite $CaSO_4$ and halite NaCl. Compositions of these minerals are not very variable.

The evaporites are rocks formed by chemical precipitation from seawater or continental waters.

Many evaporites formed by evaporation of seawater in shallow (partially) isolated basins. The order of precipitation follows the order of increasing solubility in seawater: calcite, dolomite, gypsum/anhydrite, halite, potassium and magnesium salts.

Calcium sulfates precipitate when about 30% of the initial seawater remains. The normal form of calcium sulfate in subaqueous environment is the hydrated form, gypsum $CaSO_4 \cdot 2H_2O$, that appears as well-formed 1–25 mm, often twinned, crystals, forming rosettes, stone roses, etc. In an environment subject to high evaporation (upper intertidal and supratidal environment) anhydrite may crystallize. It has a habit of fine-grained nodules, coalescing and separated by thin walls of sediment.

Halite precipitates when no more than 10% of the initial volume of seawater remains. It may be associated with magnesium sulfate (kieserite, $MgSO_4H_2O$) and chloride.

Potassium (and magnesium) chlorides and sulfates are formed when the residual water represents only 6% of the initial volume: sylvite KCl carnalite $KMgCl_3 \cdot 6H_2O$, kainite $KMgClSO_4 \cdot 3H_2O$, polyhalite $K_2 MgCa_2(SO_4)_4 \cdot 2H_2O$. There are also complex diagenetic reactions between these minerals.

The above sequence supposes a single cycle of evaporation. Evaporite deposits are generally far from being as simple in their organization, both in plane and vertically, due to the complex evolution of the basins and to new seawater inputs.

Evaporites formed in a continental environment contain halite, gypsum and anhydrite and specific minerals as sulfates and sodium carbonates. Unlike marine evaporites, potassium sulfates and chlorides are rare.

In addition to the previous minerals, continental evaporites may contain small amounts of bromides, fluorides, iodides, borates (California, Nevada, Turkey), nitrates (Chile), lithium minerals, etc. Continental evaporite contain more diverse minerals than marine evaporites because they are formed from waters that may have leached the most varied terrains – particularly rocks of volcanic origin – and/or may come from thermal springs.

Gypsum (and anhydrite) can also form in deep (at least below the limit of influence of waves), partially isolated basins. That produces euxinic conditions in the deepest parts of the basin. Gypsum and/or anhydrite formed in these conditions show a laminated facies. They are associated with carbonates and organic material.

When the burial exceeds several hundred meters, gypsum is completely transformed into anhydrite. When this anhydrite reaches the surface, by erosion or tectonics, it is rehydrated and transformed into gypsum. The habit of this secondary gypsum is characteristic; large euhedral crystals, fibrous gypsum, alabaster or massive bodies formed by small, poorly defined, intermeshed crystals.

Gypsum is a secondary mineral developed from sulphides, particularly in ore deposits. It can also be formed by reaction between the sulfuric acid formed by oxidation of sulfides with carbonate rocks, limestone or marl. It is formed in volcanic environment by the action of sulphur-bearing fumaroles on calcic minerals.

Anhydrite is a primary magmatic mineral in volcanic trachyandesitic pumices (for example in Mount Pinatubo, Philippines). As anhydrite is soluble, it is not preserved in ancient volcanic rocks.

Halite is also a deposit from fumaroles.

3.7   ORE MINERALS

Metallic minerals exploited as ore are mostly sulfides, sulfosalts, oxides and, to a lesser extent, native elements, tungstates, etc. Most of these minerals are opaque: they cannot be studied with the petrographic microscope, but require the reflected light microscope.

Some ore minerals (or associated with ores) transparent in thin section, are presented in the CD.

3.7.1   Barite

$BaSO_4$ is a common mineral in the massive sulphide deposits, veins of low temperature and stratiform deposits in carbonate (Mississippi Valley type) or clastic environment (such deposits are sometimes called Red Beds). It is very commonly associated with sphalerite, pyrite, galena and sometimes fluorite. Barite is then a gangue for these ores or it may be exploited for itself. The only significant substitution in this mineral is that of barium by strontium.

3.7.2   Fluorite

Generally contains at least 99 wt% CaF_2. The only notable substitutions are the replacement of Ca by Sr or by Ce and Y.

Fluorite is a mineral of evolved igneous rocks, both subalkaline and alkaline: granites, syenites, nepheline syenites and associated pegmatites. It is probably not primary but rather appears at the tardi-magmatic (pneumatolytic) and hydrothermal stages. It is a mineral of the greisens where it is associated with topaz, lepidolite, cassiterite, etc.. It is rare in volcanic rocks, where it occurs in geodes and fumarole deposits.

The major economic deposits of fluorite are hydrothermal veins of low to medium temperature and stratiform of Mississippi Valley type as well in carbonates as clastic environment. It is associated with sphalerite, pyrite, galena, chalcopyrite and sometimes barite. In some districts, or even some deposits, display a spatial zonation between barite and fluorite may be observed.

3.7.3   Sphalerite

ZnS is the main ore of zinc. There are currently substitution of zinc by iron; the maximum contain is about 26 wt% Fe, that is 45 mol% FeS. The association sphalerite (Zn, Fe)S, pyrrhotite $Fe_{1-x}S$, (pyrite FeS_2) can be used as a geothermometer and to determine f_{S2}. Spharerite may contain some cadmium, manganese, and trace of Ag, In, Ge, Ga, Tl, Se, Hg. These elements can greatly increase the value of the exploited sphalerite. Sphalerite

Name Chemical formula	symetry	crystalline forms - cleavages - twins position of the indicatrix	Indices 1,45 balsam 1,54 quartz 1,55 tourmaline 1,65 garnet 1,75 2,8	Color Alterations	Birefringence 0 black gray 1st order white 0,009 yellow orange color 0,018 boundary between 1st and 2nd orders 2d - 3rd orders: bright colors 0,054 higher orders: whashed colors	Uniaxial/ Biaxial +/−
Gypsum $CaSO_4 \bullet 2H_2O$	mnocl	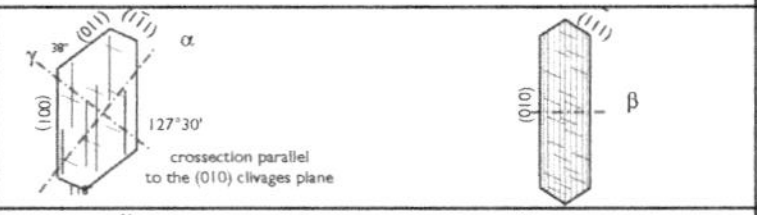(011) (111) α γ 38° (100) 127°30' crossection parallel to the (010) clivages plane (010) β	1,5205 1,5296	colorless	0,0091	B+ (2V=58°)
Anhydrite $CaSO_4$	orthor	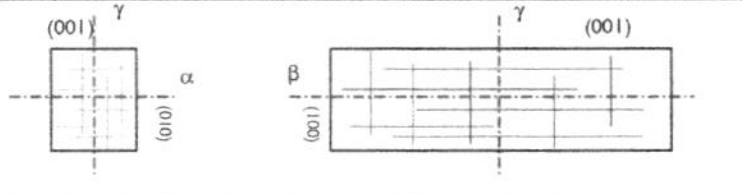(001) γ α (010) β (001) γ (001)	1,569 1,6138	colorless	0,0438	B+ (2V=42–44°)
Barytine $BaSO_4$	orthor	tabular, lamellar, radiating agregates 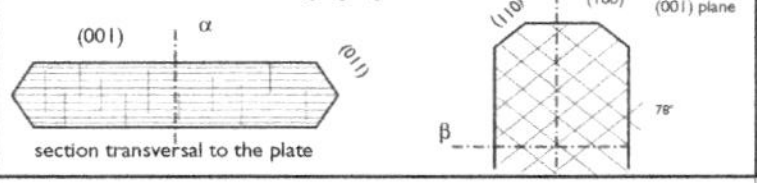(001) α (011) section transversal to the plate (110) γ (100) section along (001) plane 78° β	1,636 1,648	colorless	0,012	B+ (2V=36 −38°)
Fluorite CaF_2	cub	Octaedral (111) cleavage	1,434 negative relief !	colorless (purplish)		
Blende ZnS	cub	(110) cleavage	2,396 2,47	yellowish yellow brown red brown		
Scheelite $CaWO_3$	tetr	granular, no cleavage	1,92 1,94	colorless yellowish brownish	0,015 0,017	U+
Cassitérite Sno_2	quadr	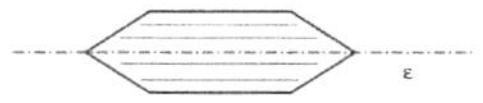ε	1,997 2,098	colorless red brown green brown zoned	0,096 0,097	U+

Figure 3.41 Summary of the characters of sedimentary and ore minerals.

frequently contains (sometimes submicroscopic) inclusions of chalcopyrite which is a result of the replacement of zinc by copper in the lattice. Sphalerite show various habits (in German sphalerite is blende and means deceiving or misleading) both at the macroscopic and microscopic scale, depending on its chemical composition (especially its content of iron, cadmium and manganese) and its more or less development of crystals: large limpid crystals, zoned concretions, microcrystalline and compact, transparent ("honey" or "ruby sphalerite") to almost opaque sphalerites.

Sphalerite is a hydrothermal mineral of medium to low temperature. It is most often associated with pyrite, pyrrhotite, galena, chalcopyrite, sometimes sulfosalts, and in some deposits, barite and/or fluorite. It is mined in many ore types: massive sulphides, skarns (especially molybdenum skarns), veins associated with Porphyry Copper, meso-to epithermal veins, stratiform deposits of the Mississippi Valley type (where it is most abundant in carbonate environment that in clastic ones).

3.7.4   Scheelite

$CaWO_4$ and wolframite $(Fe, Mn)WO_4$ are the major ores of tungsten. There is a continuous solid solution between scheelite and powellite $MoWO_4$.

Scheelite is a mineral of the skarns associated with granodiorites and monzodiorites. It appears at the main stage of skarn formation where it is accompanied by numerous calcic minerals (described in the chapter 3–4). It occurs in both hedenbergite skarns in the andradite skarn. It also appears in the late hydrothermal stages of skarns where it is associated with sulphides: arsenopyrite, pyrrhotite, etc. Scheelite also occurs in pegmatites and hydrothermal veins with wolframite (+ quartz + sulphides). Sometimes wolframite and scheelite are in reactional association, scheelite replacing wolframite or the reverse.

3.7.5   Cassiterite

SnO_2 is the main ore of tin. Iron (ferrous and ferric) can substitute to tin. Cassiterite also contains tantalum and niobium and lesser amounts Mn, Ti and Sc.

Primary deposits of cassiterite are associated with granites (s. s.), leucogranites and alkali-feldspar-granite. It may appear at the magmatic stage and tardi-magmatic stage (pegmatites with cassiterite, wolframite, topaz, beryl, etc.), but its main deposits are rather associated with the hydrothermal stage:

• greisens with lepidolite/muscovite, topaz, tourmaline, wolframite, scheelite, molybdenite, arsenopyrite, Bi–Nb–Ta;

- tin-bearing skarns (tin is often in the silicates: tin-bearing garnet, malayaïte $CaSnTiO_5$); such skarns are either calcic skarns (low in boron, rich in sulphides) or magnesian skarns (with boron minerals, pyroxenes, etc);
- stockworks of quartz veins (for example "tin porphyries" of Bolivia) with stannite, silver sulfosalts, lead sulfoantimonides, bismuth, bismuthinite;
- hydrothermal veins (for example Cornwall deposits): cassiterite, wolframite, bismuth, bismuthinite, molybdenite, chalcopyrite, pyrite, pyrrhotite, arsenopyrite, löllingite.

Cassiterite is a heavy mineral easily deposited and concentrated in placers (deposits of Malaysia).

Appendix – Calculation of the structural formula of a mineral

The chemical analysis of mineral gives the **percentage by weight** of the different oxides that compose this mineral. The analysis is presented by order of decreasing valency these oxides: SiO_2, TiO_2, Al_2O_3, Fe_2O_3, Cr_2O_3, FeO, MnO, MgO, CaO, Na_2O, K_2O, H_2O, F, Cl. The structural formula shows the distribution of the **atoms** in different sites of the unit cell (or in one of its multiple or sub multiples).

The principle of this calculation is very simple:

1   The number of atoms of each element in 100 g of the sample is obtained by dividing the percentage by weight of oxides by the molecular weight of each oxide.
2   A simple rule of three, distributes the atoms in the unit cell by fixing one or more parameters of this cell: it may be number of total number of oxygens, the number of cations in the entire or part of the unit cell, etc.

The main difficulty is the choice of this or these parameters.

When the distribution of iron between Fe_2O_3 and FeO (and manganese between Mn_2O_3 and MnO) and H_2O is known, it is usual to calculate the structural formula *on the basis of a fixed number of anions* (O, OH, F, Cl).

Chemical analysis of minerals are most often made using an electron microprobe which do not measure elements lighter than chlorine, nor H and O. Consequently, only the total iron is determined, usually in the form of ferrous iron, which we will note FeO*. It is the same for manganese. *The previous calculation based on a content of oxygen and hydroxyl can no longer be done* and can only lead to *inaccurate* results.

The method of calculation must be adapted to each particular case.

I – if we can assume that **there is no water, nor ferric iron** in the structural formula, it is possible to calculate this formula on a base on a number of oxygen or on a number of cations.

Example 1: A plagioclase $(Na, K)_{1-x} Ca_x Al_{1+x} Si_{3-x} O_8$ [albite (x = 0), anorthite (x = 1)]. Calculation of the structural formula can be based either on 8 oxygens, or on 5 cations.

Table I. Detailed example of calculation of the structural formula of a mineral without water nor without ferric iron a plagioclase (Na,K)$_{(1-x)}$Ca$_x$Al$_{(1-x)}$Si$_{(3-x)}$O$_8$.

	Weight %	Molecular weight			In 99,2 g (total weight) of the mineral		Calculated structural formula	
					Number of cations	Number of oxygens	On a base of 5 cations	On a base of 8 oxygènes
SiO$_2$	59,43	60,09	60,09	Si	0,98902	1,97803	2,68	2,67
TiO$_2$	0,03	79,90	79,90	Ti	0,00038	0,00075		
Al$_2$O$_3$	25,17	101,94	50,97	Al	0,49382	0,74073	1,34	1,33
Cr$_2$O$_3$	0	152,02	76,01	Cr	0	0		
Fe$_2$O$_3$	0	159,70	79,85	Fe^{3+}	0	0		
FeO	0,02	71,85	71,85	Fe^{2+}	0,00028	0,00028		
MnO	0,03	70,94	70,94	Mn	0,00042	0,00042		
MgO	0	40,32	40,32	Mg	0			
CaO	7,47	56,08	56,08	Ca	0,13320	0,13320	0,36	0,36
Na$_2$O	6,93	61,09	30,55	Na	0,22688	0,11344	0,61	0,61
K$_2$O	0,12	94,20	47,10	K	0,00255	0,00127	0,01	0,01
BaO	0	153,36	153,36	Ba	0	0		
F	0	19,00	19,00	F	0	0		
Cl	0	35,46	35,46	Cl	0	0		
H$_2$O	0	18,02	9,01	OH	0	0		
total	99,20			O	1,84654	2,96813	8,04	4,97
				number of cations			5,00	8
factor					2,70776	2,69530		

% An 36,7
% Ab 62,6
% FK 0,7

Remarks

1-there are two atoms of Al in 101,94 g of Al$_2$O$_3$; to obtain the number of atoms you have thus to divide by 101,94/2 = 50,97.

2-it is enough for the structural formula to round the results to two decimal digits.

3-the quality of the analysis is estimated by concordance with the theoretical structural formula, the value of the parameter x must be identical for each cation and lesser important, the total sum of weight % that should be 100%.

The main cations found in the compositions of common minerals are here indicated, even if they are not involved in this example, so that to give a table of their molecular weights.

II – Calcul of the distribution of iron between ferrous and ferric iron

1   The structural formula is calculated on a number of cations in all, or part, of the unit cell;

2   The amount of ferrous iron and ferric iron is equal to the total iron:

$$Fe^* = Fe^{2+} + Fe^{3+}$$

3   The number of oxygens corresponding to the number of cations is fixed:
= 2 (Si + Ti) + 1,5 (Al, + Fe^{3+} + Cr) + (Fe^{2+} + Mn + Mg + Ca) + 0,5 (Na + K).
So we have two equations to calculate Fe^{2+} and Fe^{3+}

4   The contents of FeO and Fe_2O_3 can then be recalculated in weight of oxides in the analysis.

Example 2: An aluminous diopside Ca (Mg, $Fe^{2+})_{1-x}$ (Al^{VI}, $Fe^{3+})_x$ Si_{2-x}, $Al^{IV}_x O_6$ number of cations = 4 = 6 number of oxygens.

Table 2. Example of a mineral with ferrous and ferric iron without water aluminous diopside Ca (Mg, $Fe^{2+})_{1-x}$ (Al^{VI}, $Fe^{3+})_x$ $Si_{2-x}Al^{IV}_x O_6$.

	Weight %					
	Results of the microprobe	Recalculated analysis	Structural formula			
SiO_2	45,76	45,76	Si	1,72		Z site (2)
TiO_2	0,96	0,96	Aliv	0,28		
Al_2O_3	8,45	8,45	total Al		0,38	
FeO*	9,77		Alvi	0,10		
Fe_2O_3		5,11	Ti	0,03		
FeO		5,18	Fe^{3+}	0,14		
MnO	0,09	0,09		R^{3+}	0,27	Y site (1)
MgO	10,15	10,15	Fe^{2+}	0,16		
CaO	24,01	24,01	Mn	0,00		
Na_2O	0,31	0,31	Mg	0,57		
K_2O	0,00	0,00		R^{2+}	0,73	
BaO	0,14	0,14	Ca	0,97		X site (1)
total	99,64	100,16	Na	0,02		
			K	0,00		
			Ba	0,00		
			Ca + Na + K + Ba	0,99		

Parameters of the calculation.
Number of cations = 4.
Number of oxygens = 6.
Columns of intermediate calculations are omitted.
Total iron = $Fe^{2+} + Fe^{3+}$ = Fe = 0,31.
Number of oxygens calculated on the previous number of cations.
Regardless the distribution of ferrous iron = 5.93.

III – Calculation the percentage of H_2O is by fixing the sum OH + F + Cl according to the theoretical structural formula. The quantities of OH in the structural formula OH and H_2O wt% in the analysis can then be recalculated.

Example 3: A biotite

Potassium in biotites (and more generally in micas) is often poorly measured, for several reasons:

- K and Na are volatile components that evaporate under the beam of the microprobe; it is preferable to make the determination of such elements early in the cycle of dosage and possibly move the sample a little after having measured them;

Table 3. Example of a mineral with water a biotite $X_2 Y_{4-6} Z_8 O_{20} (OH)_4$.

	Wt %						
	Results of the microprobe	Recalculated analysis	Structural formula				
SiO_2	39,01	39,01	Si	5,76			site Z (8)
TiO_2	1,20	1,20	Al IV	2,24			
Al_2O_3	15,78	15,78			Al total	2,75	
FeO	14,58	14,58	Al VI	0,51			
MnO	0,04	0,04	Ti	0,13			
MgO	14,96	14,96	Cr	0,00			
CaO	0,00	0,00			R^{3+}	0,64	site Y (5,74)
Na_2O	0,13	0,13	Fe	1,80			
K_2O	9,41	9,41	Mn	0,01			
F	1,32	1,32	Mg	3,29			
Cl	0,13	0,13			R^{2+}	5,10	
H_2O		3,40	Ca	0,00			
	96,56	99,96	Na	0,04			
			K	1,77			
					Ca + Na + K	1,81	site X (1,81)
			OH	3,35			OH + F + Cl = 4
			F	0,62			
			Cl	0,03			

Parameters of the calcuation.
It is assumed that all the iron is at ferrous state.
K, Na, Ca not taken into account.
Number of oxygens corresponding to the other cations = 21.
 OH + F + Cl = 4
This biotite is not strictly octahedral.
K, Na, Ca are probably under-estimated or this biotite is slightly chloritized.

Columns of intermediate calculations are omitted.

- depending on whether the beam is perpendicular or parallel to cleavage, the measures of the elements located in the interlayers may be different;
- biotites are sometimes slightly chloritized with some leaching of potassium, even if they appear optically fresh;
- it is better to avoid to take into account K, Na, Ca in the calculation of the structural formula.

Micas are not strictly dioctahedral or trioctahedral because there is some extent of solid solution between the end members. The number of cations in the cell is not known.

Table 4. Example of a mineral with ferric and ferrous iron and water a hornblende $A_{0-1}X_2Y_5Z_8O_{22}(OH)_2$.

	Wt %		Structural formula				
	Results of the microprobe	Recalculated analysis					
SiO_2	42,35	42,35	Si	6,28			Z site (8)
TiO_2	0,39	0,39	Al^{IV}	1,72			
Al_2O_3	16,25	16,25			total Al	2,84	
Cr_2O_3	0,91	0,91	Al^{VI}	1,12			
FeO*	16,55		Ti	0,04			
Fe_2O_3		3,72	Cr	0,11			
FeO		13,21	Fe^{3+}	0,41			
MnO	0,28	0,28			R^{3+}	1,68	Y site (5)
MgO	7,46	7,46	Fe^{2+}	1,64			
CaO	11,12	11,12	Mn	0,04			
Na_2O	1,70	1,70	Mg	1,65			
K_2O	0,47	0,47			R^{2+}	3,32	
BaO	0,00	0,00	Ca	1,77			A (2) and
F	0,00	0,00	Na	0,49			X (0,34) sites
Cl	0,00	0,00	K	0,09			
H_2O		2,02	Ba	0,00			
total	97,48	99,88	Ca + Na + K + Ba			2,34	
			OH	2,00			OH + F + Cl = 2
			F	0,00			
			Cl	0,00			

Parameters of the calculation.
  Number of cations others than Ca, Na, K = 13.
  Number of oxygens corresponding to the other cations = 21.
  OH + F + Cl = 2

  Total iron = Fe_{2+} + Fe_{3+} = Fe = 2,05.
  Numbers of oxygens corresponding to the previous cations without taking into account.
  The ferric-ferrous ratio = 20,793.

Columns of intermediate calculations are omitted.

It is more simple to assume that iron is mainly in the ferrous state and to calculate the corresponding number of oxygen for the cations other than K, Na, Ca (that is 21 oxygens in this part of the unit cell), regardless of OH ions that will be latter independently calculated.

Example 4: A hornblende $A_{0-1} X_2 Y_5 Z_8 O_{22} (OH)_2$

In amphiboles, the A site (occupied by Na and K) is not necessarily complete, and on the other hand, Na may also occupy the X site (Ca). It is better to ignore these elements and calculate the structural formula on the Y and Z sites of the unite cell, that is on 13 cations. The ratio ferrous – ferric iron is then recalculated from the corresponding number of oxygen (21) and OH fixing $OH + F + Cl = 4$.

If there is sodium in site X, the calculation may be erroneous. So consult Leake (1978) for calculating the structural formulas of amphiboles in function of the type of amphibole.

When we the various parameters (H_2O, FeO, Fe_2O_3) have been calculated, it is possible to recalculate the analysis in weight of the mineral, to make the total and compare the recalculated analysis the raw results of the microprobe.

The criteria of quality of analysis are:

1   Match with the structural formula.
2   To be compatible with other analysis of the same mineral in the same sample or in related samples; indeed there are regular laws of variation of the chemical composition of minerals that reflect the geological processes; the new analysis should integrate the set of analyses that describe this phenomenon. If not, it is worth reflecting on why it does not fit.
3   The total of the analysis should be 100 wt% after recalculation; this is perhaps the least important criterion; some water-rich mineral (such as chlorites), or minerals with a light matrix (cordierite with Si, Al, Mg and possibly zeolitic water) or minerals with a sharp contrast in weight between the various elements (garnet with heavy elements like Fe, Mn, and light ones, Si and Mg), give an apparently poor result; some tolerance of the total may be accepted, as long as this tolerance is of the same order for all the minerals that we want to compare.

There is not single way to calculate a structural formula. We have to adapt this calculation to the minerals – and the problems – that are studied. However, it must be clearly indicated how the calculation was done.

A selection of books

CRISTALLOGRAPHY – CRYSTALLINE OPTICS

Fischesser, R., 1962. Cours de Cristallographie. Optique des milieux anisotropes, E.N.S.M.P., 89 p.

Fischesser, R., 1971. Cours de Cristallographie. Cristallogaphie géométrique, E.N.S.M.P., 107 p.

Nesse, W.D., 2000. Optical Mineralogy. Oxford University Press, 322 p.

MINERALOGY – DETERMINATION OF THE MINERALS

Deer, W.A., Howie, R.A. and Zussman, J.: Rock Forming Minerals.

First Edition

Vol. 1 – Ortho- and Rig Silicates, Longman, 1962, 333 p.
Vol. 2 – Chain Silicates, Longman, 1963, 379 p.
Vol. 3 – Sheet silicates, Longman, 1962, 270 p.
Vol. 4 – Framework Silicates, Longman, 1963, 435 p.
Vol. 5 – Non Silicates, Longman, 1962, 371 p.

Second Edition

Orthosilicates, Geological Society of London, 1982, 932 p.
Disilicates and Ring Silicates, Geological Society of London, 1986, 630 p.
Single Chain Silicates, Geological Society of London, 1978, 680 p.
Double Chain Silicates, Geological Society of London, 1997, 784 p.
Vol. 3 A – Micas, Geological Society of London, 2006, 780 p.
Vol. 4 A – Framework Silicates – Feldspars, Geological Society of London, 2001, 992 p.
Non Silicates: Sulfates, Carbonates, Phosphates, Halides, Geological Society of London, 1995, 392 p.

Deer, W.A., Howie, R.A. and Zussman, J., 1966. An Introduction to the Rock-Forming Minerals, Longmans, 528 p.

———, 1992. Second edition, Longman Scientific and Technical, 696 p.

Fabriès, J., Touret, J. and Weisbrod, A.: Détermination des Minéraux des roches au microscope polarisant de Marcel Roubault, 4ème edition, 1982, Editions Lamarre- Poinat, 383 p.

Gaines, R.V., H. Catherine Skinner, H.C., Foord, H.C., Mason, B., and Rosenzweig, A., 1997. Dana's New Mineralogy, John Wiley & Sons, 1819 p.

Hey, M.H., 1954. A new review of the chlorites. *Mineral. Mag.*, 30: 277–292.

Leake, B.E., 1978. Nomenclature of amphiboles, *American Mineralogist*, 63, 1023–1052.

MacKenzie, W.S. and Guilford, C., 1980. Atlas of rock-forming minerals in thin section, Longman, 98 p.

Roubault, M., Fabriès, J., Touret, J. and Weibrod, A. 1963. Détermination des minéraux des roches au microscope polarisant; Lamarre-Poinat, Paris, 365 p.

Strunz, H. and Nickel, E.H., 2006. Strunz Mineralogical Tables. Chemical – Structural classification System, 9ème édition. E. Schweitzerbart'ssche Verlagbuchhandlung, Stuttgart. 870 p.

Tröger, W.E., 1971. Optische Bestimmung des gesteinsbildenden Minerale, 4. neubearbeitete Auflage von H.U. Bambauer, F. Taborszky und H.D. Trochim, Stuttgart, E. Schweizerbart'sche Verlagbuchhandlung, 188 p.

Reviews in Mineralogy and Geochemistry – Mineral Society of America – Geochemical Society

Vol. 2. Feldspar Mineralogy 2nd. ed., 1983, P.H. Ribbe, ed., 362 p.

Vol. 3. Oxyde Minerals, 1976, D. Rubble, ed., 300 p.

Vol. 5. Orthosilicates, 1980, P.H. Ribbe, ed., 450 p.

Vol. 7. Pyroxenes, 1980, C.T. Prewitt, ed., 525 p.

Vol. 9A. Amphiboles and others hydreous Pyriboles – Mineralogy, 1986, R.D. Veblen, ed., 372 p.

Vol. 9B. Amphiboles: Petrology and Experimental Phase Relations, 1982, D.R. Veblen and P.H. Ribbe, ed., 390 p.

Vol. 11. Carbonates Mineralogy and Geochemistry, 1983, R.J. Reeder, ed., 367 p.

Vol. 13. Micas, 1984, S.W. Bailey, ed., 584 p.

Vol. 19. Hydrous Phyllosilicates (Exclusive of Micas), 1988, S.W. Bailey, ed., 725 p.

Vol. 22. The Al_2SiO_5 Polymorphs, 1990, D.M. Kerrick, ed., 214 p.

Vol. 25. Oxide Minerals, 1991, D.H. Lindsley, ed., 509 p.

Vol. 46. Micas. Cristal Chemistry and Metamorphic Petrology, 2002, A. Mottana, F.P. Sassi, J.B. Thompson and S. Guggenheim, ed., 499 p.

Vol. 53. Zircon, 2003, J.M. Hanchar and P.W.O. Hoskins, ed. 500 p.

Vol. 56. Epidotes, 2004, A. Liebscher and G. Franz, ed., 628 p.

Petrography – Texture of the rocks

MacKenzie, W.S., Donaldson, C.H. and Guilford, C., 1980. Atlas of igneous rocks and their textures, Longman, 148 p.

Pizigo, M., 1965. Les principales textures de roches, Rapport interne B.R.G.M., 70 p.

Vernon, R.H., 2004. A practical guide to Rock Microstructure, Cambridge University Press, 594 p.

Petrology

Best, M.G., 1982. Igneous and Metamorphic Petrology. W.H. Freeman and Company, New York, 630 p.

Best, M.G. and Christiansen, E.H., 2001. Igneous Petrology, Blackwell, 458 p.

Bowen, N.L., 1928. Evolution of the Igneous Rocks, Princeton University Press, Princeton, 334 p.

Cox, K.G., Bell, J.D. and Pankhurst, R.J., 1979. The Interpretation of Igneous Rocks, George Allen and Unwin, 450 p.

Carmichael, I.S.E. Turner, F.G. and Verhogen, J., 1974. Igneous Petrology, McGraw-Hill Book Company, 739 p.

Henderson, P., 1982. Inorganic Geochemistry. Pergamon Press, 353 p.

Miyashiro, A., 1973. Metamorphism and Metamorphic Belts, John Wiley and Sons, New York, 492 p.

Miyashiro, A., 1994. Metamorphic Petrology. John Wiley and Sons, New York, 1994, 404 p.

Tucker, M.E., 1981. Sedimentary Petrology, Blackwell Scientific Publications, 252 p.

Winkler, H.G.F., 1979. Petrogenesis of Metamorphic Rocks. 5th Edition, Springer Verlag, New York, 1979, 348 p. (1ère edition 1965).

References

Bayliss, P., Mazzi, F., Munnio, R. and With, T.J., 1989. Mineral nomenclature: zirconolite. Mineraligocal Magazine, n°53, p. 565–569.

Boissonnas, J., 1973. Les granites à structure concentriques et quelques autres granites tardifs de la Chaîne pan-africaine en Ahaggar (Sahara central, Algérie). Thèse de Doctorat ès-Sciences Naturelles, Paris VI, p. 662 + annexes.

Bouladon, J., 1986. La chromite: un minerai toujours recherché. *Chron. Rech. Min.*, n°485, p. 53–63.

Bowen, N.L., 1915. The later stages of the evolution of the igneous rocks, *Journal of Geology*, 23, p. 1–89.

Boyd, F.R., 1959. Hydrothermal investigations of amphiboles. Researches in Geochemistry, Abelson ed., Wiley, New York, p. 377–396.

Buddington A.F., 1963. Isograds and the role of H_2O in origin of the amphibolites in the North-West Adirondacks, New York. *Geol. Soc. Am. Bull.*, 74, p. 1193–1202.

Carlson, W.D., 1988. Subsolidus phase equilibria on the forsterite saturated join $Mg_2Si_2O_6$ – $CaMgSi_2O_6$ at atmospheric pressure. *Amer. Min.*, 73, p. 232–241.

Chappell, B.J. and White, A.R.J., 1974. Two Constrasting Granite Types. *Pac. Geol.*, 8, p. 173–174.

Chopin, C., 1979. De la Vanoise au Grand Paradis. Une approche pétrographique et radiochronologique de la signification géodynamique du Métamorphisme de haute pression. Thèse 3ème cycle, Univ. Paris VI, p. 145, 5 pl. 7 tabl.

Chopin, C., 1984. Coesite and pure pyrope in high-grade blueschists of the Western Alps: a first record and some consequences *Contributions to Mineralogy and Petrology* 86: 107.

Coombs, D.S., 1960. Low grade mineral facies in New Zeland. *Internat. Geol. Congr. 21st Sess. Rep*. Part 13, 339–51. Copenhagen

Crawford, W.A. and Fyfe, W.S., 1965. Lawsonite equilibria. *Amer. J. Sci.*, 262, p. 262–270.

Demange, M., 1976. Une paragénèse à staurotide et tschermakite d'Ovala (Gabon). *Bull. Soc. fr. Minér. et Cristallogr.*, 99, p. 379–402.

Demange, M. 1976. Le métamorphisme progressif des formations d'origine pélitiques du flanc sud du massif de l'Agout (Montagne Noire, France). 2ème partie. Variations de la composition chimique des minéraux. *Bull. Minéral.*, 101, p. 350–355.

Demange, M. and Gattoni, L., 1976. Le métamorphisme progressif des formations d'origine pélitiques du flanc sud du massif de l'Agout (Montagne Noire, France). 1ére partie. Isogrades et façiès de métamorphisme. *Bull. Minéral.*, 101, p. 334–349.

Demange, M., 1982. Etude géologique du massif de l'Agout, Montagne Noire, France. Thèse d'Etat, Université Pierre et Marie Curie, Paris VI, t.1 408 p. 167 fig., t.2 p. 647.

Demange, M. and Machado, R., 1998. O batólito cordilherano Serra dos Orgãos: um exemplo de arco magmático brasiliano com assinatura toleítica no sistema de cisalhamento Paraíba do Sul, no estato do Rio de Janeiro. 38° Congresso Brasileiro de Geologia, Balneário Camboriú, SC, 1994, p. 114–115.

Demange, M., Pascal, M.-L., Raimbault, L., Armand, J., Serment, R., Forette, M.-C. and Touil, A., 2006. The Salsigne Au – As – Bi – Cu – Ag deposit, France. *Economic Geology*, 101, p. 199–234.

Dempsey, M.J., 1981. Zussmanite Stability; A Preliminary Study. *Progress in Experimental Petrology*. Vol. 5, p. 58–60.

Einaudi M.T., and Burt D.M., 1982. Introduction, terminology, classification and composition of skarn deposits. *Economic Geology*, 77, p. 745–754.

Eskola, P., 1915. On the relations between the chemical and mineralogical composition in the metamorphic rocksof the Orijärvi region. *Bull. Comm. géol.Finlande*, N°44.

Eskola, P., 1920. The mineral facies of rocks. *Norsk Geol. Tidsskr.*, 6, 143–194.

Eskola, P., 1929. On mineral facies. *Geol. Fören. Stockholm Förh.*, 51, 157–172.

Feenstra, A. and Engi, M., 1998. An experimental study of the Fe-Mn exchange between garnet and ilmenite. *Contribution to Mineralogy and Petrology*, 131, n°4, p. 379–392.

Florke, O.W., Jones, J.B. and Schmincke, H.- U., 1976. A new microcrystalline silica from Gran Canaria. *Zeitschrift für Kristalographie*, 143, 156–165.

Florke, O.W., Florke, U. and Giese, U., 1984. Moganite – a new microcrystalline silica-mineral. *Neues Jahrbuch für Mineralogie Abhandlungen*, 149, 325–336.

Folk, R.L. and Land, L.S., 1975. Mg/Ca ratio and salinity: two controls over crystallization of dolomite. *Bull. Amer. Ass. Petrol. Geol.*, 59, p. 60–68.

Fonteilles, M., 1970. Géologie des terrains métamorphiques et granitiques du massif hercynien de l'Agly (Pyrénées orientales). *Bull. Bur. Rech. Géol. Min., Orléans*, 2ème série, sect; IV, n°3, p. 1–72.

Fonteilles, M. and Guitard, G., 1971. Sur les conditions de formation du grenat almandin et de la staurotide dans les métapélites mésozonales hercyniennes des Pyrénées orientales: mise en évidence de variations mineures du type de métamorphisme. *C. R. Acad. Sci. Paris*, ser. D, 273, p. 659–662.

Fuhrman, M.L. and Lindsley, D.H., 1988. Ternary-feldspar modelling and thermometry. *American Mineralogist;* 73; no. 3–4; p. 201–215.

Gebauer, D., Williams, I.S., Compston, W. and Grünenfelder, M., 1989. The development of the European Continental Crust since the early Archean based on conventional and ion-microprobe dating of up to 3.84 by old detrital zircons. Tectonophysics, 157, p. 81–96.

Gilbert, M.C., 1966. Synthesis and stability relations of the hornblende ferropargasite. *Amer. J. Sci.*, 264, p. 698–742.

Goffé, B., Goffe-Urbano, G. et Saliot, P., 1973. Sur la présence d'une variété magnesienne de ferrocarpholite en Vanoise (Alpes françaises). *C. R. Acad. Sci. Paris*, ser. D, 277, p. 1965–1968.

Grubenmann, U, 1904–1096. Die Kristallinen Schiefer. 1rst ed. I (1904), II (1906), 2nd ed. (1910). Berlin: Gebrüder Borntrâger.

Hayob, J.L., Bohlen, S.R. and Essene, E.J. 1993. Experimental investigation and application of the equilibrium rutile + orthopyroxene = quartz + ilmenite, *Contribution to Mineralogy and Petrology*, 115, n°1, p. 18–35.

Henry, D.J. and Guidotti, C.V., 2002. Titanium in biotite from metapelitic rocks: Temperature effects, crystal-chemical controls and petrologic applications. *American Mineralogist*, 87, p. 375–382.

Henry, D.J., Guidotti, C.V. and Thomson, J.A., 2005. The Ti-saturation surface for low-to-medium pressure metapelitic biotites: Implications for geothermometry and Ti-substitution mechanisms. *American Mineralogist*, 90, p. 316–328.

Hey, M.H., 1954. A new review of the chlorites. *Min. Mag.*, vol.30, p. 277.

Ishihara, S., 1977, The magnetite-series and ilmenite-series granitic rocks: *Mining Geology*, 27, p. 293–305.

Jenkins, D.M. and Bozhilov, K.N., 2003. Stability and thermodynamic properties of ferro-actinolite: Are-investigation. *Amer. J. Sci.*, 303, p. 723–752.

Kitahara, S., Takeneuchi, S. and Kennedy, G.C., 1966. Phase relations in the system $MgO - SiO_2 - H_2O$ at high temperatures and pressures. *Amer. J. Sci.*, 264, p. 223–233.

Kushiro, I., 1969. The system diopside – silica with and without water at high pressures. *Amer. J. Sci.*, 267 A, p. 269–294.

Lindsley, D.H., MacGregor, I.C. and Turnock, A.C., 1964. Synthesis and stability of ferrosilite. *Carnegie Inst. Whashington, Ann. Rept. Dir. Geophys. Lab.*, p. 148–150.

Liou, J.G., 1971. P–T stabilities of laumontite, wairakite, lawsonite and related minerals in the system $CaAl_2Si_2O_8 - SiO_2 - H_2O$. *J. Petr.*, 12, p. 379–411.

Metz, P, 1970. Experimentalle Untersuchung der Metamorphose von kieselig dolomitischen Sedimenten. II. die Bildungsbedingungen des Diopsids. *Contribution to Mineralogy and Petrology*, 28, p. 221–250.

Michel-Lévy, A., 1894–1904. Etude sur la détermination des feldspaths dans les plaques minces au point de vue de la classification des roches. Librairie Polytechnique Baudry et Cie, Paris, 108 p. + 18 p.

Muir Wood, R., 1972. The iron-rich blueschist facies minerals. I- deerite. Mineral. Mag., vol. 43, p. 251–259.

Mukhopadhyay, A., Bhattacharga, A. and Mohanty, L., 1992. Geothermometers involving clinopyroxene, garnet, plagioclase, ilmenite, rutile, sphene and quartz: estimation of pressure in quartz-absent assemblages. *Contribution to Mineralogy and Petrology*, 110, n°2–3, p. 346–354.

Newton, R.C. and Kennedy, G.C., 1963. Some equilibrium reactions in the join $CaAl_2Si_2O_8 - H_2O$. *J. Geophys. Res.*, 68, p. 2967–2984.

Nickel, E.H., 1995. The definition of a mineral. *The Canadian Mineralogist*, 33, p. 689–690.

Pascal, M.-L., 2005. Relics of high temperature clinopyroxene on the join Di-CaTs with up to 72 mol.% $Ca(Al, Fe^{3+})AlSiO_6$ in the skarns of Ciclova and

Magureaua Vatei, Carpathian, Romania. *The Canadian Mineralogist*, vol. 43, p. 857–881.

Parson, I, 1978. Alkali feldspar: which solvus? *Physics and Chemistry of Minerals*, Springer Verlag, 2, n°3, p. 199–213.

Perkins, D. III, Westrum, E.F. Jr. and Essene, E.J., 1980. The thermodynamic properties and phase relations of some minerals in the system $CaO - Al_2O_3 - SiO_2 - H_2O$. *Geochem. Cosmochem. Acta*, 44, p. 61–84.

Pupin, J.P. et Turco, H., 1972. Une typologie originale du zircon accessoire. Bull. Soc. Minéral. Cristallogr., 95, p. 348–359.

Shannon, R.D.,1976. Revised effective ionic radii and systematic studies of interatomic in halides and chalcogenides. *Acta Cryst.* A, 32, p. 751–767 rayons ioniques.

Shannon, R.D. and Prewitt, C.T., 1969. Effective ionic radii in oxides and fluorides. *Acta Cryst.* B, 25, p. 925–946.

Shannon, R.D. and Prewitt, C.T., 1970. Revised values of effective ionic radii. *Acta Cryst.*, 26, p. 1046–1048.

Skippen, G.B., 1974. An experimental model for low pressure metamorphism of siliceous dolomitic marble. *Amer. J. Sci.*, 474, p. 487–500.

Slaughter, J., Kerrick, D.M. and Wall, V.J., 1975. Experimental and thermodynamic study of equilibria in the system $CaO - MgO - SiO_2 - H_2O - CO_2$. *Amer. J. Sci.*, 275, p. 143–162.

Thompson, A.B., 1957. The graphical analysis of mineral assemblages in pelitic schists. *Amer. Miner., Washington*, 42, p. 842–858.

Touil, A, 1994. Géochimie et minéralogie comparées d'associations magmatiques acide-basiques de type magnesiopotassique et calco-alcalin: exemple du massif de l'Agly (Pyrénées orientales). Thèse Ecole Nationale Supérieure de Mines de Paris. 502 p.

Touret, J., 1977. The significance of fluid inclusions in metamorphic rocks. In Thermodynamics in Geology, Fraser, D.G. ed. Dordrecht, Reidel Pub. Co, p. 203–227.

Touret, J., 1981. Fluid inclusions in high grade metamorphic rocks. In Short couse in fluid inclusions: application to petrology, Hollister, L.S., Crawford, M.L., Min. Assoc. Canada, Calgary, p. 182–208.

Touret, J., 2006. De la petrographie à la pétrologie. *Travaux du comité français d'histoire de la géologie (COFRHIGEO)*, troisième série, t. XX, n°8.

Tuttle, O.F. and Bowen, N.L., 1958. Origin of granite in the light of experimental studies in the system $NaAlSi_3O_8 - KAlSi_3O_8 - SiO_2 - H_2O$. *Mem. Geol. Soc. Amer.*, n°74.

Tilley, C.E. and Harwoord, H.F., 1931. The dolerite-calk contact at Scawt Hill, Co. Antrim. The production of basic alkali-rocks by the assimilation of limestaone by basaltic magma. *The Mineralogical Magazine and Journal of the Mineralogical Society*, n°132, vol. XXII, p. 439–468.

Von Gruenenwaldt, G., Sharpe, M.R. and Hatton, C.J., 1985. The Bushveld complex. Introduction and review. *Economic Geol.*, vol. 80, p. 803–812.

Willemse, J., 1969. The geology of the Bushveld complex, the largest repository of magmatic ore deposits of the word.

Websites

www.webmineral.com
www.mindat.org
www.webmineral.brgm.fr

Subject index

T - #0027 - 160425 - C0 - 234/156/12 [14] - CB - 9780415684217 - Gloss Lamination